JOURNEYS END

By

John Brage

Journeys End - The Journeyers' Tale Book 3
Copyright © 2020 by: John Brage
Written by: John Brage

Book Cover design by: Keith Draws https://keithdraws.wordpress.com/

ISBN: 978-0-578-76632-4

10 9 8 7 6 5 4 3 2 1

Published by: Journeycraft Books

Printed and Published in the U.S.A.

For my wife; she is my tether.

Acknowledgments

So many people have contributed either directly or indirectly to this book. My family has always been so supportive of my efforts. My friend, Dr. G, has always been so helpful with everything from deep "sciencey" stuff to "what about when this happened on page.......?" But most of all, I'd like to say thanks to my readers. You have kept me going and I hope that this offering satisfies your curiosity about the direction this trilogy has taken.

And as always, happy reading!

But afterwards there occurred violent earthquakes and floods; and in a single day and night of misfortune all your warlike men in a body sank into the earth, and the island of Atlantis in like manner disappeared in the depths of the sea. For which reason the sea in those parts is impassable and impenetrable, because there is a shoal of mud in the way; and this was caused by the subsidence of the island.

— Plato, Timaeus, 360 B.C.

CHAPTER I

"M-my Lord," gasped the Guard. "Please! I need to rest." The huge Umae staggered along behind Dom. They hadn't stopped at all for nearly four full rotations. Now in the middle of a broad, grassy plain, the Guard collapsed. Dom turned and looked down at him, stone-faced.

"You managed to die more slowly than the others," he noted. "Not that it's any consequence." The full moon had risen well above the horizon and its face stared down at them. Dom's lips curled into a hungry sneer. "Min…." he growled. "How I hope that you aren't dead. How I hope that you are alive and happy so when I find you, I can peel away every last layer of your contentment. Your misery will ooze from your pores and I will feast on it. When I'm done, you won't want to be happy ever again. You will want only death, but that I will deny you as well."

The Guard groaned and failed in his attempt to rise. "I---I," he mumbled weakly.

"Shut up and die already," ordered Dom. "Somehow the Umae and their pathetic protectors managed to destroy my Queen. They have destroyed my only opportunity to establish my dominion over this galaxy. So," he added casually as he crushed the Guard's skull with a stomp of his foot, "I'll dominate this planet. I'll dominate it soon, and I'll dominate it forever. And someday when Min's friends return, I'll immerse them in what I've created and they will drown in it."

That Guard had been the last of eighteen Dom had managed to gather after the Great Rain. One by one they had grown sick and died. It was a puzzle Dom couldn't solve although it added to his frustrations.

He wiped his foot off in the grass before peeling off his Chronicler robe and tossing it aside, leaving him clad in only his smallclothes. "There is plenty

1

of time," he said with a smile. "Plenty of time for me to create misery without measure."

Dom continued walking. He wanted to gain a better understanding of how this world had been changed by the pulse. The daytime sky was now mostly a light blue with wispy gray-white clouds. Occasionally the clouds would coalesce into dark gray masses that would pound the ground below with torrents of rain. But then the clouds would move by and the bright yellow sphere in the sky would return. He relished the sensations it brought him. He would strip completely bare and lie down to soak up its radiant energy. It was maddeningly intoxicating. His purpose was to increase the energy levels on this planet much, much more than the pulse wave had accomplished. With his Queen, he would have been able to do so and then populate the galaxy with his progeny. Now that his Queen was gone, the sunlight left him insatiably aroused. He had no outlet. The sun was simply a daily torment reminding him of what he had lost.

The night brought a different torture. As the rotations passed, the moon would gradually reveal its entire face to him in a rough twenty-eight rotation cycle. He quickly knew where on the horizon it would rise, and where it would set, and he would see both because he never slept. That face conjured images of Min in his mind. Min, the one who had destroyed his Queen and brought him these agonies. Min would learn anguish. Any suffering he had ever previously experienced in his meaningless life would pale in comparison. One night, just as the full moon had lifted itself above the horizon, Dom sat down for the first time in several scores of rotations. He closed his eyes and focused on that hated face overhead. Before, he had stretched out his consciousness in an effort to join with his Queen. That was no longer possible. Instead, he allowed his mind to return to the place where he had communed with her before. The target of his concentration was the moon. He felt himself rise from the ground and into the sky. The air was cool in the absence of the sun. He sensed the air grow thinner as he drew nearer to the edge of the atmosphere. He quickly made the passage across the expanse of space from the last thin cloud to the dusty surface of the moon itself. He sensed nothing living. Drawn towards the hum of machinery, he entered the same huge structure he had entered before, the place where he had so briefly tasted his Queen. He felt no sensation – but wait! He did feel something, the barest spark of life. Before he had sensed the radiant power of his Queen's intellect and the relative mote of neurological energy offered by Min. His Queen's blazing brilliance had been extinguished.

But Min's feeble ember continued somehow. He was faded and failing but he was still there. He was on the moon and he was alive. And he was going to pay for what he had done.

Something was wrong. The Directors had intended to increase the amount of radiant energy on this planet. Even with the clouds significantly reduced, this was nowhere near what they had intended. And it was too soon. It was supposed to be the residents of the planet that created the new environment. During his melding with his Queen, she had explained to him her plan. She had found a way to turn something Min was doing against the planet itself. Min must have interfered with that as well. So many crimes he would be held accountable for!

Dom drew his consciousness back to his body. He stood up and marched off across the plain. He desperately needed technology. The pulse had destroyed Atla. Dom understood the purpose of the Directors. He had seen it when he touched the panel in the Chroniclers' Hall. But their plans had now gone awry and that loss gnawed at him. He knew he wouldn't find any Umae settlements. They had all been destroyed. But he would find Umae. Weak, stupid Umae. They would do his bidding just as they had before. And once they had given him what he needed he would exact his revenge.

Dom had a perfect understanding of the movements of the objects in the night sky. He held a special hatred for the moon. The day before its full, taunting face would appear above the horizon, Dom would bathe his naked form in sunlight for the entire day. Then he would seethe. His one opportunity at galactic domination had somehow been stolen from him by an idiot originating from this planet of idiots. All of this pent-up energy would be devoted to Min's misery. And if his misery somehow ended before Dom was spent, then Dom would vent the remainder on the Umae. His vengeance could only be fully satisfied if he could get to where Min was on the moon. The Journeyers were his only means of doing that. He knew they would eventually return and that he had hundreds of ellipses to execute his plan.

He roamed the land far and wide, never needing to rest. The planet was very different now. The sun crossed the sky every rotation, its light and heat mostly unimpeded by the clouds. For a period, the air would grow progressively cooler each rotation. During another period, it would grow gradually warmer. During some periods, it rained more than others. There were no discrete patterns, but simple trends that he easily comprehended. Dom attributed these changes to the changing orientation of Uma to the yellow sun.

He occasionally saw traveling groups of Umae and Hek. They were together now. The Umae had lost their fear of these primitives. They would need something new to fear. Dom would provide that. These groups never remained in one place. They would arrive in an area, stay there long enough to hunt and gather plants, and then they would move on. He was never in any danger of being discovered. His senses were so keen that he could study them from a great distance without any concern about being found. But they would find him eventually. At a time of his choosing, they would find him. And then he would execute his plan. Until then, he would continue watching and learning. He would know the ways of his prey, and then he would pounce on them unsuspectingly.

There weren't nearly as many Umae as he would have expected. And his travels revealed a low ratio of young Umae to adults. They were not reproducing as quickly as he thought they did. There were no young Hek at all. He could not recall seeing one that could have been born prior to the pulse. This was something he would have to consider. Could the pulse be connected to the disappearance of young Hek and the relative infertility of the Umae? And a third group appeared. They were not Hek nor were they Umae. They were tall like the Umae but well-built like the Hek and they appeared in all age ranges. The two species were interbreeding. That wouldn't do. But they were the least of his concerns for the moment. The Umae were his future and the future of the planet. While all others would perish, the Umae would learn to regret their survival.

The sun. The sun. The blazing ball of flame poured its energies down upon Dom as he lay, spread eagle and naked, on a barren patch of ground. Dom's inner fire, that which drove him to reproduce, scorched his core. He spent an entire cloudless rotation basking, longing, planning. Then, as the light slowly faded in the eveningside sky, he waited patiently for the appearance of Min's hated face on the horizon. He knew exactly where it would rise. Tonight, its face would be full and Dom's rage would be an inferno. If he could not have the release he preferred, then he would seek out the Umae, or the Hek, and vent his frustrations upon them. Tonight, his plans would take flight. As the barest edge of the moon peeked over the planet's edge, Dom rose and sprinted off across a broad plain, leaving his coverings behind. Effortlessly he covered the ground between himself and a distant range of low mountains. He would find them there. He was sure of it.

An angry growl pierced the air. Dom stopped, tilting his head to better determine the location of its source. He began running again once he had determined the proper course. Swiftly, quietly, he reached the edge of the plain as it gave way to an expanse of scrubby bushes. Beyond that, the mountains rose. It was there that he would find the source of the sound. He saw four figures, backed up against a steep rock rise, confronted by the creator of the noise he had heard. It was a huge beast, as tall at the shoulder as the people it threatened. Two men bravely waved spears at the monster as two children cowered behind them. As one man thrust his spear forward, the creature slapped it away with a massive paw before pouncing on its attacker. The man wailed as the beast's long, curved incisors sank into his shoulder. The beast shook the man in its teeth like a toy before slinging him aside. The children cried in anguish as the remaining man boldly repositioned himself in their defense. Mouth half open, the monster assessed the situation waiting to see what his prey would do next.

"Hold!" called Dom. The spear-bearer stole a look over the monster's shoulder, disbelieving what he saw as Dom approached from behind. The huge cat paid him no mind, still focused on the man and the two children. "I said HOLD!" repeated Dom. The volume of his voice was such that it reverberated off the stone rise, amplified into the voice of a hundred men. The beast turned and hissed at him, its eyes drawn back. It crept closer towards Dom, a defiant naked child. Then it slowly lowered itself into a crouch, ready to pounce on its new target. They locked eyes. Dom's unblinking glare bore into the creature. It hissed again, refusing to look away. Then finally, it broke. Slowly moving sideways to create space between itself and this strange creature, the giant cat finally could take no more. It tore its eyes away and sprinted off into the bush, a bawling mew left in its wake.

The figure with the spear was just barely a man. He had no more than sixteen or seventeen ellipses. He was an Umae. The children peeking out from behind him were both Hek. One of them ran over to the cat's victim a short ways away. Their defender was now motionless, his blood pooling beneath him. He was a stoutly built Hek. The younger man joined the child at the fallen man's side, carefully checking his companion. After a moment, he looked at the children with a heaviness in his eyes. After he grunted something to them the little ones wailed and threw themselves down onto the dead man's form.

Dom watched them closely, careful to listen to any sounds they might make. He waited. When they finally decided to move, he would go with them. The monster was gone, but a new one had taken its place.

CHAPTER 2

Alone aboard the ***Aurora,*** Jack stood waiting for his repose chamber to open. In a few hundred ellipses he would awaken and assist his fellow Journeyers in training their students. Eve had determined that it would be better to keep everyone else on the ***Starshine.*** It was technically more advanced than the ***Aurora*** and, in her estimation, safer. Someone had to stay aboard the ***Aurora*** in case some sort of technical issue arose. With Min's absence, Jack was their foremost technical expert. But now he was something else as well.

As the door slid all the way open Jack turned to view the rest of the ship. The ***Aurora***'s Alpha panel was directly ahead of him about twenty paces away. He walked to the Command Agent's chair immediately in front of the panel and sat down.

"What is the name of this ship?" he asked, glancing up at the ceiling.

"This Journeycraft is called the '***Aurora***'." The ship's 'voice' was cold and metallic.

"I want to see the ship's statistical performance parameters. I want to see all efficiency ratios relating to propulsion, sensory capabilities, and computation." He sat back in the chair and watched the display in front of him. Almost instantly a report containing the requested information began to stream in front of him faster than he could comprehend. "Stop!" he barked. "First I just want a standard propulsion analysis. I want to know this ship's limitations. Can you manage that?"

Without bothering to answer, the ship removed the previous report from the display and replaced it with a much shorter one. It contained data related to the ship's engines. Jack leaned in and looked it over closely. He guffawed slightly, an expression of annoyance sweeping over his face. "Moons, who maintained this mess?" he asked quietly. "The ***Starshine*** has exceeded these specifications for a few hundred ellipses according to its service logs." He tapped his thumbnail against one of his front teeth. "Let me see the parameters on the communication systems," he demanded. The ship quickly complied. Once again Jack was disappointed.

"Wait a......." he began. "Tell me again, what is the name of this ship?"

"This Journeycraft is called the '***Aurora***'". A smile slowly crept up on Jack's face.

"And tell me, who is the Command Agent of the *Aurora*?"

"Jack is."

He was beaming now. "Tell me. Say 'Jack is the Command Agent of the *Aurora*.'"

"Jack is the Command Agent of the *Aurora*."

"Again!"

"Jack is the Command Agent of the *Aurora*."

"One more time!" By now he was shouting.

"Jack is the Command Agent of the *Aurora*."

He sat back in the chair and folded his arms in front of him. "Yes. Yes, I am," he said with satisfaction. "I am the Command Agent of the *Aurora*." He turned and looked back towards the repose chamber. "Reseal the repose chamber," he said. "I'm not ready for it yet."

Eve's first sensation was the burning in her nostrils. The lining of her nose stung and she sniffled in an attempt to end the discomfort. A bright light filled her vision and she could sense her eyes watering to the point of tears. She blinked a couple of times and a single tear rolled down her cheek. She was still inside her repose chamber.

In front of her was a display. It currently read 'eighty-three percent' with the number gradually rising. The tingling in her extremities soon stopped and, aside from the customary stiffness in her muscles, she felt fine. The repose chamber door slid open and Eve stepped onto the deck of the *Starshine*.

None of the other repose chambers were in the process of awakening the crew member inside. "Status?" she asked, licking her dry lips.

"The *Starshine* is receiving a message," replied the ship.

Eve rolled her shoulders, trying to loosen them. "A message? From whom?"

"It appears to be coming from the *Aurora*."

Jack. She tried to come up with as many reasons as she could as to why he would be trying to message the *Starshine*. None of them were good. "How long have I been in repose?"

"Four hundred eighty-four point six eight ellipses," replied the ship.

She walked to the Command Agent station and sat down. "Almost time to wake up anyway," she noted under her breath. "Crew status?"

"No irregularities noted. Would you like to terminate repose?"

"No, not yet. I want to find out what Jack wants first. How far ahead of us was he when he sent the message?"

"At the time of transmission, the *Aurora* was approximately .21 average planetary distances behind the *Starshine*."

"'Behind'?" she echoed, a hint of confusion in her voice. "Are we using the same orientation reference? Where was he in relation to the *Starshine* and Uma?"

"The *Aurora* was in between the *Starshine* and Uma."

"That's impossible," said Eve. She was beginning to fear that the *Starshine*'s Alpha panel was malfunctioning. "Was the *Aurora* inside the tether beam?"

"Affirmative."

"Then how…. wait, never mind." She felt a lump in her throat and her eyes began to water much more heavily than before. "Implement Min simulation." She could barely whisper his name through the tightness in her throat. She knew what was going to happen, but she couldn't prepare herself for it.

"Hello Eve, can I assist you with something?" The algorithmic approximation of her love sounded exactly like him. She covered her face as sobs racked her body. "Eve? Are you well?"

She took a deep breath and tried to gather herself. After a moment, she sat back in her chair and closed her eyes. "Can you determine how it is that the *Aurora* sent us a message when it was behind us in the tether?" That simple question robbed her of her breath and she released a new batch of sobs.

"One moment." Min had always been careful to analyze everything thoroughly before offering an opinion. His simulation was apparently designed to do the same. "It should not be possible," Min said finally. "I went ahead and calculated the position of the *Aurora* based on the vector of the message. The *Aurora* is within the tether beam. However, the tether beam can only transmit data in one direction. In this case, the *Aurora* should not have been able to send the *Starshine* a message because their orientations should not have permitted it."

Eve sniffled in an attempt to prevent her nose from running. "So how did this happen?"

"One moment." Eve detected a glitch in this approximation of Min. It had used that phrase twice now and she was sure she had never heard him say it. "The logs indicate that the *Aurora* and the *Starshine* have remained inside the tether since they left Uma and that the *Aurora* was in between the *Starshine* and Uma when the message was sent."

"So, it shouldn't have been possible for Jack to have sent the message, is that right?"

"Based on the information I have available, that is correct."

"Is there any danger in reviewing the message?" asked Eve. "I'm sorry….." That phrase caused her to unconsciously draw her arms around her chest. "But, could a message cause some functional impairment to a Journeycraft?"

"Highly unlikely. The *Starshine* is much more advanced technologically than the *Aurora*. If you'd like, I will scan the message for abnormalities before playing it."

"Yes, do that."

"One moment."

Eve bristled. "And stop using that phrase, 'one moment'. Min never says… he never said that."

"Of course." Eve realized she had been hugging herself so she reluctantly released her arms. "No threat detected. Shall I play it?"

"Yes." Eve could hear the ship's Data Station, her former position, processing the message for playback.

"Eve. There is something we need to discuss." It sounded like Jack, but she thought he sounded ill. "I'm providing you with the coordinates of a small moon in a nearby system. Meet me there as soon as you can. I need to explain in person."

"Was that Jack?" she asked the panel.

"The vocal projections were well within the ranges of those we have on file for him," replied Min. "Although somewhat to one end of the expected range in terms of tonal quality."

"I want to send a message back to him," said Eve. "And this time I want to include a visual element."

"That's not possible."

"Why not?" asked Eve. "You said he was behind us."

"He was behind us," clarified sim-Min. "But he isn't anymore. The *Aurora* has recently accelerated and moved past the *Starshine* in the tether beam. We have no way to reply."

"Except in person," said Eve.

"Correct."

Eve closed her eyes. "All right. Set a course to those coordinates. Continue long range scans on the *Aurora*. And go ahead and end repose for the rest of the crew. We have training to do anyway."

"Affirmative."

"And Min?"

"Yes, Eve?"

She stood up and steadied herself by placing a hand on the back of her chair. "Never mind."

Once her repose tube opened, Lise practically bounded forward onto the deck of the **Starshine**. Eve couldn't recall ever seeing anyone shake off the effects of repose so quickly.

Lise walked over to Mac's tube, a smile rising on her face. He was still blinking his eyes inside the tube in an attempt to get his bearings. "Where are we?" she asked Eve.

"That's a hard question for me to answer in a way you would understand," she replied. "We are a long, long ways away from Uma." Lise took Mac's hand as he stepped forward. Lin was also exiting her tube. Once Lin had gained her bearings, she assisted Frell in exiting his tube as well. The little boy smiled up at her familiar face and quickly took her hand.

"I don't feel any different," said Lise as she wrapped her arms around Mac's neck. The young man stood there, mostly impassive, although he did gently hug her in return.

"You aren't any different," said Eve. "At least not much different." She had moved over to Sol's chamber, checking the readings on the outside of his tube. His eyes were still closed.

"Is everything all right?" asked Lin.

"Yes," said Eve confidently. "Some people adjust more slowly to revival than others. And Sol......" She hesitated as she smiled at the old man in front of her. "Well, he has quite a number of ellipses."

"How long have we been gone?" asked Mac. His question caught Eve off-guard.

"According to the panel," Eve said, "right at four hundred eighty-five ellipses".

Lise's hand went to her chest. "Four hundred eighty-five? We are that much older than we were before?"

Eve was distracted now. Sol's eyes were still closed and he wasn't moving yet.

"Eve?" asked Lin quietly. She took a couple of slow steps in the direction of Sol's chamber.

Sol's eyes finally fluttered open and his body was immediately wracked by a hacking cough. He reached out in an attempt to steady himself against the side of the tube.

"Mech, I need assistance with Sol's extraction," said Eve, a new tone of concern weighing on her words. She quickly began entering directives into a small panel by his chamber as one of the **Starshine**'s mechs rolled into view. It was approximately waist high and it extended two long metal arms out in front of it, sliding their blunt ends under Sol's armpits. Sol leaned forward allowing the mech to support his weight. He was still coughing vigorously. "Sol? Can you hear me?" Eve took him by the hand.

The Life Agent took a couple of forced breaths and his coughing eased somewhat. He nodded his head as the mech helped him step out of the chamber. It immediately oriented itself so that Sol could sit on top of it. "Exam.....circlets," he gasped.

Eve pushed his long white hair away from his face. His usual pale white skin was flushed from his efforts. "Auxiliary mech, bring some examination circlets," she barked. Quickly, a smaller mech roughly a third the size of the first rolled into view, offering Eve a pair of round, silver hoops at the end of its utility arm. Eve took them and passed them to Sol. He carefully passed his hands through the circlets, sliding them midway up his forearms. As he did so, a clear membrane stretched itself out over his hands and forearms. Once the devices were in place, the membranes became so clear that they were impossible to see. Sol took a deep breath to steady himself and held his palms against his cheeks.

"What is he doing?" asked Lise. Lin gave her a sharp glance, causing the younger girl to recoil slightly. Sol's face was a rosy mask of concentration as the membranes gathered information about his physical condition and passed it to his nervous system. After a moment, he removed his hands and took another deep breath. Frell stood by Lin, watching quietly.

"Panel, direct the Life Lab to execute directive package two, five, gamma," he said finally. "Advise as to how long it will take to complete."

"The entire process will be complete in .14 rotations," responded the panel.

"Well?" said Eve, somewhat relieved that his coughing had stopped and that he was able to examine himself.

"Nothing," smiled Sol wearily. "Nothing I can't deal with anyway." He set his feet on the floor and attempted to stand.

"No!" Eve objected. "Wait. Why don't you wait?"

Sol eased himself back onto the top of the mech. "I need to get to the Life Lab."

"Clarify," answered the panel. "Was that a request for transportation assistance?"

He groaned slightly. "It wasn't," he smirked, "but I guess it is now."

11

The mech adjusted its configuration to provide him with a backrest and to assure that he was safely secured to the machine. Then it slowly started to roll in the direction of the Life Lab.

"Can I go with you?" asked Lin. Sol silently waved her along as the mech bore him down a nearby corridor. Lin followed. She released Frell's hand but the boy followed along behind her, nonetheless.

Once Lin was out of sight, Lise spoke up. "Is he all right?"

Eve wasn't sure how to answer. After all, Sol had over a thousand ellipses now. "Yes," she said finally. "And he is getting the best care possible."

Lise leaned against Mac. "What do we do now?"

"You study," said Eve. "But first you will all be evaluated. One of you will become a Life Agent, one a Data Agent, and one a Tech Agent. We will decide the course for the other one of you based on the evaluations. But first we need to determine who is best suited for which position. Once we've done that, most of your training will be done on the tutorial panels, although Sol, Jack and I will be supervising you."

"Jack," replied Lise. "Where's Jack?"

"He's still on the *Aurora*," said Eve. "But we are going to be seeing him soon. He wants us to meet in a nearby system. Apparently he has something he needs to communicate to me directly. After that, whoever will be starting Technical Agent training will go with him on the *Aurora*. The rest of you will stay here."

Lise's eyes grew wide as she turned and looked up at Mac. "How long will our training last?" she asked.

"Assuming average progress, about fifteen ellipses."

"Fifteen!" echoed Lise. She wrapped both of her arms around Mac's midsection and squeezed him hard. "But......."

"There are no 'buts'," said Eve coolly. "We might be the last chance to save the Umae from extinction. Being a Journeyer is about sacrifice. Get used to it." She barely managed to croak out her last few words.

Lise closed her eyes and pressed her cheek against Mac's chest. He responded with a one-armed hug. Eve saw the connection between the two young Umae, even though Mac was rather guarded with his displays of affection. Lise had enough for both of them. It wasn't fair. No one knew that better than Eve.

"I'm going to go check on Sol," she said finally. "I'd enjoy your time together."

As Eve, Lise and Mac entered the Life Lab, Sol was reclining on an elevated cushion raised to a 45-degree angle. Lin was next to him studying an array of containers on a tray in front of him. Frell sat at a low table examining a small, clear vial.

"How are you feeling?" asked Eve as she approached his bed. Lise and Mac trailed along behind her.

"I'm fine," said Sol. "Lin has been very helpful."

"'Fine'," said Eve. "Tell me what 'fine' is."

Sol smiled. "Eve, it isn't uncommon for......more seasoned...travelers to have difficulties when ending a repose cycle." Catching the concern rising on her face, he added, "And no, I'm not talking about RTS. There are a myriad of other, less serious conditions, that can manifest much in the same way that it does though. I have nothing more than a bit of a dry cough and a common upper respiratory infection."

There were a number of metallic tables about the room bearing a variety of different devices relating to Sol's position. While Lise and Mac seemed mostly interested in listening to what Sol had to say, Lin eagerly took in as much detail of the room's contents as she could.

"So how long?" asked Eve.

"How long until what?" replied Sol with a grin.

"Until you can assume full duties. We need to get the training assessments done. And Jack wants to meet with us."

Sol adjusted himself on his cushion. "Less than a rotation," he said. "I've already started the process of assembling the proper phage to dispose of the intruders in my lungs. A container of water should take care of my other problem."

"I can get that for you," said Lin. "I mean, if you tell me where it is."

"Don't worry," said Sol, patting her hand, "I can have the mech handle it." He turned back towards Eve. "Jack wants to meet with us?"

Eve seemed uncomfortable. "I think it would be best if you waited for me back on the main bridge," she said to Lin, Lise, and Mac. "We won't be long." The three youths complied, exiting through the door they had entered earlier. Frell looked up and upon seeing Lin leaving with the others, scrambled after her. With Eve's command, the door slid shut. "Jack sent us a message," said Eve. "But it is rather strange."

"Strange? How so?"

"Because according to the panel and......Min's analytic profile.....Jack sent us the message while he was behind us in the tether beam."

Sol scowled. "I didn't think that was possible." He began coughing again. "Mech, bring me some water please."

Eve placed a hand on his shoulder until he stopped coughing. She couldn't help but notice how frail he felt. "Are you sure.....?"

"Yes," Sol insisted. "I am. Now, what about Jack's message?" The mech had arrived bearing a container of water. Sol accepted it and began drinking.

"He wants to meet us on a nearby moon. I wanted to send him a reply, but after sending us his message, he accelerated past us and is now ahead of us in the tether beam. We have no way to respond. So, we will have to see him in person."

"You are sure there is no way to reply?" asked Sol.

Eve's face clouded over. "Min.....I mean, the Min simulation confirmed it. According to it, there is no known method to send a message to another Journeycraft at that range without using the tether."

"But Jack did somehow," noted Sol.

"Yes."

Sol set his water aside and leaned back. "That is odd. No indication about what he wanted to talk about?"

Eve shook her head. "None."

"Well, what choice do we have? How far away is this moon?"

"It should only take us about sixteen rotations to get there. The timing of Jack's message was almost ideal."

"What do you mean?"

"The arrival of the message was what caused my repose cycle to end. We weren't due to end the cycle for another fifty-four rotations. Jack would have had to scan the expanse of space ahead of us to find a suitable meeting place and then time the arrival of the message, however he sent it, so that it arrived just as we were nearing that location."

"I'm not an expert of hard technology," said Sol, "but I don't remember Hab or Min ever doing that sort of scan. What are your thoughts?"

"I don't remember anything like that either," said Eve. "But on the other hand, we never had a need to find a specific spot like that. Our scans are extremely broad and are only searching for very general readings."

"A planet that might match the requirements for Haven," added Sol.

"Exactly. Not to say that the scan couldn't be done, just that we have never had any reason to do it." Eve's composure slipped slightly. "Min was always talking about how he wanted to focus on technology that we would actually use, not just technology for technology's sake."

Sol took her by the hand. "And he learned that from Hab," he said warmly. "But maybe the Tech Agents from the *Aurora* didn't share that philosophy." He cleared his throat. "But enough of that for the moment. You know you don't have to use Min's simulation if you need technical advice. There are many others that are available."

Eve nodded slowly. "I know that. It's just…well, I needed to hear his voice. He's……."

"He's what?"

She paused and removed her hand from his. "He was the best, right?"

"The best Technical Agent, you mean?" Now Sol was smiling broadly. "I suppose he still is. And I still have all of my data from my RTS trials. I'm not done working on that yet. I plan on making sure he is the best for a long time yet."

Eve turned and squeezed her eyes closed tightly, trying to keep the tears from returning.

"We can get the assessments done before we reach this moon, can't we?" asked Sol. "Whoever my Life Agent student is going to be will be helping me with my research. Don't be surprised if that turns out to be Lin."

"Why do you say that?" asked Eve.

"She has 'the touch'," answered Sol. "Knowing all the science behind the Life Agent's duties is one thing, but she has that extra bit of empathy that I think is necessary for the job. The others may be plenty bright, but they are pretty fixed on one another still. I fully expect Lin to be the one to help me find a way to get Min out of that repose tube at the Citadel." Sol chuckled. "Young Frell is certainly attached to her. I'm curious as to how his assessment will go."

"Oh? Why so?" asked Eve.

Sol took one more sip of water before setting his water container aside. "It's my understanding that he is non-verbal. That doesn't mean he isn't intelligent. It could mean any number of things. His assessment will give us more insight into his neuro-psychological status."

"He's non-verbal? How did you know that?"

"Lin told me before we lifted off. Apparently Bal didn't think that was something we needed to know. He's barely above the age for the commencement of Journeyer training. Lin was beside herself over the prospect of leaving him behind. Bal may have been trying to get rid of him."

"He may well see her as a mother figure," said Eve. "Even though May was highly regarded as a Rearer, perhaps Frell needed a bit more individual attention."

Sol nodded. "Exactly my thoughts. After all, May was attempting to raise her own biological child. No matter how hard she tried I'm sure it was difficult for her not to favor Dom. But I'm sure he will be fine. Regardless of his assessment results, we can find something for him."

Eve nodded slightly before continuing towards the door. "I hope you are right," she said quietly as the door slid open in front of her.

Once she had exited, Sol coughed for a moment before considering the collection of vials in front of him. He hoped that he had caught his own RTS early enough that its advance could be severely impaired. The renewal of his research wasn't just an attempt to save his friend. Sol wanted very much to live long enough to celebrate the salvation of the Umae as well.

Lin, Lise and Mac sat nervously in a row watching Eve closely. Frell fidgeted next to Lin. Sol sat in a chair off to the side of the room. Eve was reviewing the data from their aptitude testing for at least the ninth time. The young Journeyers anxiously awaited the announcement as to which of them would be studying what.

"Eve?" Lise was easily the most nervous of the three.

"Yes, Lise."

"Once we get our assignments, where will we be studying?"

Eve sat back in her chair. "You will go wherever your teacher is," she said simply. "Sol and I will remain on the **Starshine**, so the training for the new Data Agent and the new Life Agent will take place here. Jack will be on the **Aurora**, at least for the time being, so the future Technical Agent will be there with him."

"For the time being?" asked Mac. His question caught Sol's attention as the young man rarely spoke. His eyes met Eve's.

"I misspoke," said Eve. "He will be on the **Aurora** indefinitely. I don't foresee a set of circumstances that would change that."

Lise spoke carefully. "Do you know?" She reached over and took Mac's hand.

"I do," said Eve. She cleared her throat as the three students all leaned slightly forward. Frell watched but seemed disinterested. "Lise, you will train with me to become a Data Agent."

Lise turned towards Mac, her eyes beginning to well. "And……..?"

Eve took a moment to check her own composure. She didn't want to crack in front of the students. "Lin, you will be studying with Sol to become a Life Agent."

16

"No!" bellowed Lise as she seized Mac's arm and tried to bury her face in it. Sobs shook her slight form as Mac tried to comfort her. He whispered quietly in her ear but it didn't seem to do any good.

"I'm sorry Lise and Mac," said Eve steadily. "But this is the optimal training arrangement based on our evaluation. It isn't as if you will never see each other again. And once we are finished training, we will have to provide both ships with full crews. I can't promise you anything, but perhaps the arrangements will change once we reach that point."

Lise looked up, her pale pink eyes rimmed with red. "But you said the training takes fifteen ellipses! Fifteen! I can't wait that long!"

Sol stood up and walked over to where the students were seated. "Lise," he said calmly, placing a hand on her shoulder, "Journeyers have always made sacrifices for the betterment of the Umae. We all long for a normal life at some point. But we don't get to do that. The best we get are glimpses of how things might be different." He noticed Eve biting her lower lip.

"What about Frell?" asked Lin as she placed a protective arm around his shoulder. He leaned in closer to her.

Eve and Sol exchanged glances. "He didn't participate in his evaluations," said Sol. "Not exactly."

Cued by Lin's obvious confusion, Eve interjected. "He had sections where he plainly understood what he was supposed to do and actually did exceptionally well. But in the other sections, his responses were mainly.......nonsensical. Statistically speaking, it was as if he were just playing a game of some sort. A game in which he was the only one who knew the rules."

Lin knelt down next to the boy. "Frell? Why did you do that? I told you how important those tests were. I thought you were going to try?"

Frell's eyes widened as he beamed at Lin. He put his arms around her neck and tried to give her a hug. Lin hesitantly returned his embrace.

"It's fine," said Eve. "Perhaps we can test him later. He's very young for Journeyer training anyway. Maybe with more ellipses he'll gain more maturity and we can get a better understanding of his talents."

"He's very smart," said Lin quietly as she extracted herself from his embrace. "He just......sees things in his own way. He's been like that since he was old enough to really interact with the rest of us."

"Nothing wrong with idiosyncrasy," said Sol as he grinned at Frell. "Variation is an important key to life. All life. Like Eve said, we will find a place for him. It may just take some time."

"When do we start?" asked Lin. She stood up and placed an arm around

Lise's other shoulder. "It will be fine," she said quietly. "You'll see. Remember what Cap and May used to tell us? 'Things are never as good or as bad as you think they are going to be'."

Lise lowered her head. "Look what happened to them."

"But they wouldn't want you sitting here feeling sorry for yourself," continued Lin. "You know that."

"We will meet with Jack before we begin training," said Eve. "That will give us the chance to move Mac to the *Aurora*."

"How long?" Lise almost sounded as if she were being strangled.

"Two rotations. We won't require any training between now and then. The three of you can do whatever you'd like until then." Lise responded by embracing Mac with both arms and squeezing him as tightly as she could.

"Eve?" said Sol. "A word in the Life Lab?"

Eve nodded and the pair exited the room, leaving the students to deal with the news of their assignments.

Eve didn't have occasion to visit the Life Lab very often. It was Sol's domain and, unless someone was ill, she simply wasn't needed there. It looked to her exactly as it had when she had spoken to Sol about his illness. Since then he had demonstrated steady improvement and now appeared to be in the same condition as when they all last left Uma. Various devices were running experiments of some unknown nature, but her only concern as it applied to Sol was his health.

The Life Agent sat down in a large chair in front of the Lab's Alpha panel. Eve sat down nearby.

"Are you sure you made the right decision?" he asked.

"About the assignments? Yes. Given the test results, they were the logical assignments."

Sol studied her face for a long time before responding. "Logic? Nonsense. Everything doesn't have to be logical. You could have just as easily sent Lin to the *Aurora*. Would that have been such a problem?"

"Lin scored the highest in all three areas," said Eve defensively. "But Mac's technical aptitudes were slightly higher than Lise's."

"'Slightly,'" repeated Sol. "But you said there was no statistically meaningful difference."

"But if we sent Lise to the *Aurora*, they still wouldn't be together," noted Eve. "So why does it matter?"

"They care about each other very, very much," said Sol. "In terms of future performance, assigning Lin to Jack would make very little difference at all, but it would let Lise and Mac be together. In your calculations, was that considered to be a benefit or a burden to future performance?"

Eve stood up. "I don't have to have this discussion with you," she said firmly. "I've made my decision."

Sol nodded. "You have, but can't you answer the question of an old friend?"

Eve seemed to relax a bit. "Very well. You are right. It wouldn't make much, if any, of a statistical difference if I assigned them differently. But I'm going to be training Lise, so I can help her deal with it."

The Life Agent frowned. "Deal with what?" he asked, perplexed.

Eve's composure was beginning to crack. "The fact that......" Tears began rolling down her cheeks. "That.....you just can't have that as a Journeyer. I did her a favor. She will never know that kind of hurt. She may feel it a little at first, but she will get over it. If they are together now, we can't guarantee they always will be. We don't know what might happen."

"You think you've lost Min forever, don't you?" asked Sol sadly. "That we won't be able to bring him out of repose when we return?"

Eve chose not to speak as the tears dripped from her chin. She simply shook her head.

"Well, I'll take that as a challenge," said Sol, "not as an assessment of my talents." The old man grinned warmly.

Eve turned quickly and left the Life Lab, heading back to her position on the bridge.

CHAPTER 3

"Begin landing sequence," ordered Eve.

"Beginning landing sequence." The Hab simulation's voice sounded exactly as she remembered. On Sol's suggestion, she had placed the navigation of the *Starshine* in that simulation's hands as opposed to sim-Min's. The Journeycraft began to lower itself slowly towards the small moon that hung precisely at the coordinates Jack had said it would.

The moon's mother planet was a medium-sized mass of swirling gasses. While it had an impressive diameter, it had no solid surface so the gravity well it created wasn't problematic for the ship at all. This moon was rather far away from its parent. Jack had chosen the meeting place very well.

The moon itself was an orange-brown chunk of rock. It had very little atmosphere and had been bombarded by space debris for millions of ellipses. Jack's message had contained the location of a specific impact crater for their reunion. As the *Starshine* cleared the outer edge of the crater, they could see the *Aurora* already resting at the bottom.

"There he is," noted Sol. "Just like he said."

Eve fidgeted while she studied her display. "But what does he want?" she asked. "This feels so strange."

The three older students were picking up on Eve's discomfort. Frell was staring intently at his own lap. "He didn't say what he wanted?" asked Lin.

"Just that he wanted to meet and he had something to tell me," said Eve.

"So, why didn't he just tell you?" asked Lin. "That is strange."

"It's not that easy," began Eve. "But we will have to explain it later. Right now," she said, focusing her attention on the panel, "we need to land."

Sim-Hab gently lowered the Journeycraft to the bottom of the crater about 500 paces from the other ship. Once the landing sequence was complete, Eve sat focused on the display while the others waited quietly.

"Hab, I want you to analyze him as soon as he comes out," she said. "I want to make sure everything is fine before we lower the egress ramp."

"That won't do any good," said the sim-Hab.

"Why not?"

"He will be in a vacuum suit," replied the panel. "It provides almost complete protection against radiation from the outside of the suit. Our scans won't be able to penetrate it."

Eve began wringing her hands. "Sol? How do we know it's him?"

Sol's left eye cocked. "Who else would it be?" he asked.

"I meant, how will we know something isn't wrong with him. This just doesn't feel right."

"From the standpoint of a Life Agent, I can't imagine what might be wrong with him. Knowing Jack, he's probably just trying to show off. Min never let him do that, you know."

Eve frowned and Sol immediately regretted his last comment. "Then hail the *Aurora*. We can communicate from this range."

"Initiating." The image of the *Aurora* vanished from Eve's display and was replaced by a blank gray rectangle. After a moment, the rectangle also vanished and was replaced by a figure wearing a vacuum suit, its head already encased in a solid black spherical helmet.

"Hello Eve," said the figure. "I'm ready to come over."

Eve listened closely to his voice. It did sound like Jack. Almost. "Hold on." She reached out and cut off the audio component of their connection. "Hab, please do a voice analysis. Compare it to the data we have on Jack."

"Analysis complete. Vocal patterns are nearly identical. The tonal quality of his voice is approximately eighteen percent lower than what our records indicate."

"Why?" asked Eve. She looked at Sol. "Why would that be?"

"Maybe he has a cold," suggested Sol. "Or maybe it's the helmet."

Eve re-engaged the audio. "Jack, can you take off your helmet?"

The figure didn't move for a moment, the black helmet making it impossible to see the face beneath. "Take it off? Why?" Eve now perceived what sim-Hab had said about the depth of his voice.

Eve glanced over at Sol. Lin, Mac and Lise shared glances with one another, trying to figure out if any of them understood what was going on better than another. Frell held a tiny bottle of brightly colored liquid. He was busy shaking it and watching the hue within make the transition from one color of the spectrum to the next. "We want to see you," she added, trying to smile.

"I want to see you, too," said Jack. "Which is why I'm coming over. I've got something exciting to tell you."

"Jack, no......" The communication was terminated by the *Aurora*. Her screen was blank.

"The egress ramp on the *Aurora* is lowering," announced sim-Hab. The video of the ship appeared on Eve's display. The other crew members of the *Starshine* moved in behind her to see better. Once the egress ramp had opened all the way they could see a figure in a vacuum suit emerge and begin walking across the moon's surface towards the *Starshine*.

"Gather as much data as you can, Hab," implored Eve. "Gait length, arm swing, whatever. All of that data is in memory."

"We have no data involving Jack walking in a vacuum suit," explained sim-Hab plainly. "And there is no statistical basis for me to make analytical adjustments for that in this environment."

They all watched in silence as the figure quickly closed the space between the two ships.

"Eve?" said Sol finally. "We should let him in, don't you think?"

Eve stared at the display for another moment. "I suppose. Lower the egress ramp and let him board."

The hum of the egress ramp ran through the bridge area briefly before a second hum indicated that it was shutting once again. They could hear the sound of approaching footsteps. The doorway to the bridge slid open and a vacuum-suited figure stepped in. It reached up and began to remove its helmet.

As the figure leaned down to set its helmet on the floor, Eve took a deep breath when she saw a cascade of long cream and white hair pour out of the black sphere. As it stood up, it wiped the long hair away from its face.

"Did you miss me?" It was Jack, except it wasn't Jack. His hair was long, nearly to the middle of his back. Its cream color was beginning to fade into a subtler hue of white. As he smiled, Eve noted the lines at the corners of his eyes.

"Jack?" asked Sol slowly. "What have you been doing?" He walked over and gave the man a hug. Jack stood there and endured it momentarily before stepping away.

"Research," he said proudly. "And have I got some things to share."

Sol stepped slowly away and gave Jack a visual once over. "What happened? You….."

Jack raised a hand to the old Life Agent, bidding his silence. "My ship, the *Aurora*, is the finest Journeycraft ever known to our people," he said flatly. "It is now capable of many wonders, both done and undone."

Sol glanced at Eve. "'Your' ship?" asked Eve.

"Yes. I am the Command Agent of the *Aurora*. I considered changing its name but decided against it. It requires no crew. Only me."

"Jack," continued Eve, "we are crewing the *Aurora*. I don't care what improvements you have made. We have a planet to find and possibly modify. That will require both ships to be fully crewed. And......" she added, waving at the Wards, "we have trainees to educate. We need your help to do that."

Jack studied the Wards for a moment. "Which one is mine?" he asked curiously.

"Which.....?" Eve was momentarily at a loss for words. "None of them are 'yours'. But Mac is going to be your Tech Agent trainee." The slender, quiet boy swallowed nervously as Lise clung to his arm.

"Him?" asked Jack. "Tell me, what were your qualification scores? You had better be smart, because I won't have time to train someone who isn't." He glared at Mac. "Understood?"

Mac nodded, almost imperceptibly.

"Jack," said Sol once again. "What did you do with the *Aurora*?"

An enormous grin blossomed on Jack's face. The others were surprised that he didn't burst out into laughter.

"Well, none of you are able to fully appreciate the technical aspects," he began, "but rest assured that its sensors, propulsion system, and computational capabilities are far beyond any ever seen on a Journeycraft before. That's no idle boast. I checked the available data. The *Aurora* is five-fold the ship the *Starshine* is."

Sol considered Jack's reply. "But the *Starshine*'s specifications were achieved primarily through Hab and Min's research. I know. I watched them do it."

Jack rolled his eyes. "And they were perfectly.....capable....technicians. But I...... Well, you will have to accept my word. The *Aurora* is a marvel."

"You had something you wanted to tell me," said Eve. "What is it?"

"You had better sit down," offered Jack. "I have a lot to say." Eve and the others found seats and waited for Jack to begin. He seemed to revel in the attention. "When we were attempting to find Max's work schedule by tapping into the Alpha panel on Atla, we had to send information through the tether beam in the wrong direction. While there was a risk of creating a feedback wave, I figured out a way to avoid that. But during that process I noted what I thought were a few irregularities in the tether itself. As we had more important matters to attend I didn't spend any time analyzing it. But once I was on the *Aurora* I had quite a lot of extra time." He stopped talking, looking back and forth at Sol and Eve.

"And?" asked Eve finally.

"In short, the reason that my workaround was successful is because there is a secondary tether beam that runs parallel to the primary one that conveys information in the opposite direction. Instead of delivering information to Uma from space, it does just the opposite."

Eve folded her arms, her eyebrows tight. "If that's true, why didn't we discover it before?"

"We never had a reason to look," said Jack. "We were instructed that the tether only transferred information from space to Uma. The Chroniclers couldn't contact us, but we could contact them. And since Journeycraft always traveled alone, there was never a need for Journeycraft to have a method of communicating back and forth. That was what I was working on when I discovered this other tether, a way for the *Aurora* and the *Starshine* to communicate directly with one another without relying on the tether and irrespective of their orientations to one another. The technology is childlike but we never had a need for it so we never bothered implementing it."

"Why weren't we told of this other tether?" asked Sol.

"The Chroniclers," said Eve. "All communications had to go through them. By the Protocols of the Directors." Eve stood up and began to pace as she tended to do when working on a problem. "So everything, every bit of information that ever went through any of the Chronicler Halls on the entire planet, was beamed into space?"

Jack nodded. "Every last bit."

"To where? To whom?"

Jack took a deep breath and allowed the suspense to build. "As I mentioned, the sensors on my ship are much more powerful than those on yours. Both tether beams change directions at an angle of roughly 32.64 degrees in approximately thirteen light ellipses."

"They bend?" asked Eve.

"No. More like a ricochet," said Jack. "My guess is that they are being intercepted by something and redirected."

"By what?

"I only have a theory." Once again, Jack fell silent.

"Well?!?" asked Eve, exasperated. "What is it?"

"A beacon satellite," said Jack finally. "It's the only instrument known to Umae that redirects a tether."

The weight of his response only fell fully on Eve and Sol. "How did a beacon satellite get so far away from Uma?" asked Sol.

"The satellite wouldn't need to travel that far," said Jack. "Only the technology to build the satellite."

"So another Journeycraft made it out this far and left a beacon satellite in the middle of space?"

"That assumes that the technology for the beacon satellite came from Uma and went into space," noted Jack.

"Wait a moment here, Jack!" insisted Eve. "What are you getting at? That something out there built a beacon satellite without any input from us?"

"I'm saying that we shouldn't assume that the technology originated on Uma, with us."

"Then.......who?" asked Eve finally.

"The Directors," said Jack.

Sol stood up and walked over next to Eve, placing a gentle hand on her shoulder. "It makes sense. We have no information about The Cataclysm. We have always assumed that the Umae developed technologically without interference from the outside. But that assumption could be wrong. Suppose the Directors, may they guide our way, established a baseline technological level for us before they disappeared? And left only The Protocols to guide us?"

"The Directors," repeated Eve. "The Directors built this beacon satellite and they knew how to do it because.......because they were the ones who taught us how to build them in the first place."

"Exactly," said Jack.

"So these tether beams, they are deflected. Why? Deflected where?"

"I told you that the sensors on the *Aurora* have been radically improved," said Jack. "I've been conducting scans looking for evidence of the planet every Journeycraft in our race's history has been looking for. "I've also tracked the tether for as far as the *Aurora* is able."

"And?"

"There is compelling evidence that a planet matching the proper characteristics is in a system directly on the path of the deflected tether beam. Eve, Sol.......I believe the tether beam will lead us directly to Haven, and quite possibly, to the Directors themselves."

"You have clearly given this a lot of thought," said Sol. "How long did you work on all of these projects before entering repose?"

Jack stood next to one of the *Starshine*'s sensor panels. "How did they not think of this?" he asked. "Once you see it, it's rather obvious."

Sol glanced over at Eve. The length of their friendship enabled her to note the subtle distress in his expression. "Jack? Sol asked you a question. How long did you stay outside of repose?"

"I want you to see the *Aurora*," said Jack. "Or at least I want Sol and Eve to see it. The rest of you won't be able to remotely appreciate the changes I've made to it." He pointed at Mac. "And you, too. If you are going to be my student, I'll want to give you a personalized tour. It will help you understand my expectations for you."

Mac, for once, shrank closely against Lise.

"I'll have to view it another time," said Sol. "Perhaps once we are closer to…..wherever it is we are looking for now. Besides, I have some work in the Life Lab that I need to attend." He stood up and gestured towards Lin. "And since YOU are MY student, perhaps you'd care to join me?"

Lin smiled and moved to his side. "I'd like that very much," she replied. "Frell? Want to come?" Frell was already at her elbow, ready to depart.

"Very well, so Eve, why don't you and…….." Jack stared at the young man clinging to Lise's arm.

"'Mac'," said Lise. "His name is 'Mac'."

"Ah yes. Mac. Why don't you and Mac suit up and join me on the **Aurora**. I think you will find it highly enlightening."

"But we need to talk some more about your theories," protested Eve. "And we need to come up with a plan of action. Together."

"Of course we do," said Jack. "But to do that, I think you'll need an appreciation for the capabilities of my ship. And besides, I thought you might want to review the data I've collected. After all, you are the closest thing we have to a Data Agent."

Eve scowled, but she didn't let her anger get the best of her. "Fine. Lise, I have arranged for some fundamental tutorials to be available on the Data panel. Why don't you begin going over those. Mac and I will go to the **Aurora** with Jack. Sol?"

The old man nodded. "We Life Agents will just do what we do," he said. He coughed slightly and tried to clear his throat. "Don't get lost."

"But……" Lise was reluctant to release Mac's arm as he started towards Jack.

"No 'buts'," said Eve firmly. "Do it. Mac, come with me and I'll show you how to use a vacuum suit."

Lise finally released Mac and he headed off with Eve towards a storage area. Lise watched them go until they rounded a corner at the far end of the hall before turning towards the Data panel.

After a short walk, Eve opened a storage chamber and removed two vacuum suits. "This is just like getting dressed," she told Mac. "Mostly. Put your feet in first and then slide your arms into the sleeves. I'll do the rest."

By the time Eve had completed the process of donning her suit, Mac had only made minimal progress. His feet were securely in the boots built into the lower section of the suit but he was having difficulties manipulating some of the controls required to get his upper body inside. Eve, her face now concealed inside the suit's jet-black helmet, stood by watching him.

"Um......" Mac said uncomfortably.

Eve reached out, took his hand, and slid it into one sleeve. After pulling the shoulder of the suit on over Mac's shoulder, she repeated the process on the other side. She made a couple of minor adjustments to the suit's controls and Mac easily shrugged the suit on so both shoulders fit perfectly. Once he realized how the suit was supposed to fit, his face blazed bright red.

"Don't worry, Mac." Eve's voice was soothing despite the hollowness caused by the helmet she was wearing. "Just learn a little bit every day. You'll get there."

Once she had secured the body of the suit, she retrieved the helmet from the storage chamber and placed it over Mac's head. She gave it a slight turn and a hissing noise indicated the correct seal.

Mac took a deep breath. Although the helmet was jet black, his vision was just as it had been moments earlier without the helmet. He looked down at his hands, watching his fingers flex inside the suit's gauntlets. "This is..... amazing," he muttered.

"It just gets more so," said Eve. "Come on." She started back down the passage to where Jack was awaiting them. He had already donned his suit and was waiting by the ship's egress portal. Without a word, he activated the portal upon seeing his approaching companions. "Follow us," said Eve.

Jack led the way down the ramp followed by Eve and Mac. Mac walked carefully for fear of falling. He found that walking in the suit was essentially the same as walking without it. While it looked large and bulky, it was actually lightweight and manageable. "This is.....amazing," he offered again.

Eve allowed herself to laugh at his choice of words. She fondly remembered her early days on the *Starshine*. While the training was exceptionally difficult, the wonders and insights it afforded were, indeed, amazing. And adding to all of that amazement was the time she got to use getting to know Min better. Now she was glad that her helmet concealed her face.

"Come on!" demanded Jack. "You will want to see this." He took long strides and quickly began to cover the distance to the *Aurora*. Eve waved Mac forward and then followed along behind him.

By the time Eve and Mac reached the base of the other ship Jack had already lowered its egress ramp. He ushered them both upwards into its belly. Eve noted the absence of a decompression wall.

"How do we take our suits off?" she asked.

"There is an E-M force fence in place," said Jack. He was studying the wrist of his suit. "And it is functioning perfectly." He reached up and promptly removed his helmet, despite the opening to the moon's surface just below them.

Mac turned towards Eve, awaiting direction. "Don't do anything yet," she said. "I'll go first." Hesitantly, she removed her helmet as well. Fresh air filled her nose and she could feel the interior warmth of the ship on her skin. "Explain that to me later, Jack," she said. She reached over and helped Mac remove his helmet. After studying the area at the top of the ramp, he looked down at the bleak, barren dust beneath them. Subconsciously, he took a couple of short steps back away from the top of the ramp.

"Now, please join me on my ship," said Jack. A section of wall slid aside and Eve could see the bridge beyond. She tucked her helmet under one arm and moved forward, followed by Mac. The sight of the bridge was almost overwhelming.

The panels were all at least half the size of those on the *Starshine*. The displays were lit with many more bright, vibrant colors and provided images cast in three dimensions in the areas nearby. From a nearby hallway emerged a tall, Umae-shaped machine. It walked forwards and stopped next to Jack. Eve estimated that it was at least thirty percent taller than he was and indeterminately more massive. Its hands had essentially the same structure as her own, albeit much larger. Its body was covered with a Journeyer uniform. Eve found slight relief in the trainee's badge it bore indicating the rank Jack had assigned to it.

"What is this?" Eve asked simply.

"A mech!" exclaimed Jack. "No more of those clumsy units on rollers. My mechs are much nimbler, much more versatile and have many, many times the processing capacity as those old ones. And I only had to build this one. Then I taught it how to build the others."

Eve didn't see any other such units on the bridge. "How many others are there?"

"Six," said Jack. "Which is actually three more than I need. But producing them was effortless, at least for me, so I decided to have extras available. Certainly, you will want some on the *Starshine*."

"Not just yet," objected Eve. "I'm going to need to review their AI parameters for starters. Mechs producing mechs? You understand the potential problems that might cause?"

"Nonsense!" snapped Jack. His prior enthusiasm was swept away by personal offense. "I reviewed them myself. In fact, I developed them." He seemed intent on boring a hole in Eve with his glare.

"All the more reason to have them reviewed by a D.A.," she replied. "Developer's bias? I'm sure Min taught you about that."

Jack scoffed. "Min….." he muttered.

"What did you say?"

Jack forced a smile. "Nothing. Of course, he did. But feel free to review them. I'm sure you will be impressed by my work. It's unprecedented. I checked."

Eve took another long look around the bridge. "Let's hope. Now, you had some data you wanted me to review?"

Jack placed his hand on the Alpha panel's display and closed his eyes briefly. "There, I've given you access to my information logs. The data I've collected about the tether, its deflection, and all the other things we've talked about is available to you now."

"You had this information on lock out?" asked Eve. "Why?"

Jack spread his arms wide. "Because this is my ship," he answered quickly. "I am the Command Agent of the *Aurora*."

A metallic voice sounded from overhead. "Jack is the Command Agent of the *Aurora*."

A much more familiar voice filled Eve's ears, nearly causing her knees to buckle.

"Jack is the Command Agent of the *Aurora*." It was Min. More specifically, it was Jack's mech speaking with Min's voice.

Eve took a deep breath through her nose. "Do all of your mechs sound like…..that?" she asked finally.

Jack nodded. "I thought it was fitting."

Eve blinked hard and refocused. "All right, I'll review the data, but it might take me a bit."

"That's fine," said Jack. His prior enthusiasm was returning. He took Mac by the arm. "That will give me time to show my student around. What was your name again?"

"'Mac'," mumbled the young man.

"Yes. 'Mac'. Come along, Mac. You have a chance to achieve greatness. If you don't, it won't be my fault."

Mac turned towards Eve. "Go ahead, Mac. It will give you something constructive to do with your time. But Jack, please try and remember that he hasn't even started his training yet."

Jack pointed at a passageway at the far side of the room. "We will start down there," he said. Jack started in that direction and Mac followed along. The gigantic mech remained.

"You are leaving this thing here?" asked Eve.

"In case you need help," said Jack, not bothering to stop.

Eve waited until the pair were out of sight before looking up at the mech. "Fine, just don't talk to me." She pressed her hand to the Alpha panel's display and closed her eyes. Her connection to the panel was nearly instantaneous — much quicker than she had ever experienced. She had no problem locating Jack's data and slowly pulling it into her own mind. Once she finished, she opened her eyes. Jack and Mac had not returned and the mech had not moved. "I wonder what access I have?" she said quietly. Once more, she joined her nervous system with that of the ship and began a search. The log entries, all of them since the time they had left Uma. They were all there. *Well Jack,* she thought as her mind probed the panel, *let's find out what all you have been up to.*

CHAPTER 4

Dom was growing impatient. The young man and the two children would not leave the fallen Hek. The children knelt next to him, sobbing quietly. The Umae man clutched his spear, carefully observing their surroundings and allowing the boys to release their grief. He would occasionally peek at Dom with an expression of awe. Finally, Dom approached the two little boys and tugged at them, trying to force them to move. Each time they bawled in protest before struggling wildly against his grip.

"They must grieve," said the young man. Dom's impatience was pressing him to simply rip their arms off and beat them to death, but he was ready for the next phase of his plan. He needed these people, at least for now. Dom sat down in the grass and waited.

Eventually, the man approached him, pointing at the waning sunlight above the eveningside horizon. "It is time for us to move on. You may come with us if you choose." He then turned towards the boys and began speaking to them in Hek. Dom listened closely to this rough tongue. He knew what the man was trying to express. Dom was more interested in how he was expressing it. He also listened to the murmurs and quieting cries of the boys near the body. Indeed, there was a pattern.

"Must go," grunted Dom at the boys. He articulated the same idea to the boys that the man had conveyed to him and with very similar sounds. The boys became quiet, awash in surprise. Their eyes searched the other man for answers but he too was stunned. "Dark soon," grunted Dom. "Must go."

"Yes, we must go," the man said in Umae. "But we can't leave his body here," he added with a gesture towards the corpse. He said something to the boys that Dom could not quite interpret. The boys reluctantly tore themselves away from the body and moved over next to the man. He provided each boy with a quick hug.

"Why not?" asked Dom impatiently. "It is useless now."

The man glanced down at the children. "Alitus was their father," he explained, "and an honored member of our clan. We must observe the Acts of Death at the main camp."

"I will bear him," said Dom.

The man paused to determine if Dom was being serious. "You

can't," he said finally. He is much too big. We will construct a litter and drag him."

"There is no time," noted Dom. "You said so yourself." He approached the body and easily lifted it off of the ground, eliciting surprised gasps from the others. The man hesitated before motioning the boys forward. Dom followed along behind them.

As the group walked along the base of a rise, the man conversed frequently with the boys. Dom was focused on every syllable in every exchange. Once the remains of the rotation had vanished, the man and the two boys stopped.

"Are you cold?" asked the man. "You have no clothes."

The boys immediately began looking around for kindling while the man began arranging some rocks into a ring on the ground. Dom stood by content for now to simply observe.

"No, I'm not cold," replied Dom. "How much farther?"

"How much farther to what?" replied the man.

"How much farther until we are with the rest of your people?"

The man finished arranging the stones for the fire ring. "You know of my people?" he asked suspiciously.

Dom nodded. "I know a great many things," he said confidently. "I have come to help you. And your people."

The boys returned, each with a small arm load of sticks. They dropped them into the circle. The man knelt down and piled them into a pyramid shape. "This won't last very long," he noted. "Did you two see any larger pieces nearby?"

Both boys shook their heads. "We didn't want to go too far," one admitted. "Not without him," he added with a subtle nod towards Dom.

"We must brave the Sky Fire," said the man. "With just the four of us and only one warrior the night is too dangerous. Alitus may draw predators. We will be careful to find shade as we can."

The boys appeared dismayed by this prospect.

"You are safe with me," Dom insisted. "I won't let anything hurt you."

The man stood up and lifted his spear from the ground. "Just who are you?" he said finally. "You are no older than they are and you have no clothes. No weapons. No food. How can this be?"

The boys were riveted now, awaiting a reply.

Dom stood up as tall as his spindly form would allow. "I was taught to protect my people, all people. I can provide you with a great many gifts. Do not be misled by my stature. I have walked this ground for far, far longer than you. Be afraid no longer, for now a Journeyer walks among you."

The man noticeably stiffened causing the boys to retreat behind him. "B-but……they are gone, never to return," he stammered. "How?"

Dom pointed to the sky, a new inspiration rising inside his mind. "See?" he asked, gesturing at the circular face beaming down on them from just over the horizon. "We never left. While some of us went into the sky, I remained here. That face is our promise to the people of Uma. A promise not to abandon you in these new times. You likely don't remember what it was like before the storm."

The man shook his head. "I've too few ellipses," he admitted. "But I know the tales of the Great Storm."

"The storm was not the end," insisted Dom. "But the beginning. The beginning of a new age. An age in which I will lead you and your people to greatness. I've no need of clothes, or food, or weapons. Despite my size, the beast who killed your friend feared my might, did you not see?"

"I saw."

"Yes. You saw." He turned towards the boys. "So, tell me your names." Dom had easily offered his request in Hek.

The first boy took a half-step forward. "I'm Clagen," he said. "And this is Tybor."

"And you may call me 'Dom'," said Dom. "And what of you?" he asked, turning towards the man.

"It seems you speak Hek as well," he noted. "I've had many names," he said. "Depending on who calls me. But you can call me by my birth name. You may call me 'Joba'".

"Joba….." repeated Dom. "Well met."

Joba and the two boys slept in relative comfort as Dom kept watch and made sure the fire didn't die. These menial duties aggravated him greatly, but they were essential to his overall scheme. As the first sign of light arose on the morningside horizon, Dom fidgeted, waiting for them to awaken. Joba was the first to do so.

"How did you sleep?" Joba asked. He noted that Dom was standing in the exact same spot as when Joba had fallen asleep the night before.

"I didn't," said Dom. "I require very little sleep."

Joba sat up and moved over to Alitus's corpse. "We need to get him to the main camp as soon as possible."

"The main camp," repeated Dom. "How far away is it? Do you have many people there?"

"Not far now. We should easily make it back before the Sky Fire flees." Joba nudged the two boys, rousing them awake.

Dom studied the morningside horizon. The sky was nearly cloud-free and the moon had set long ago. "How many are in your clan?" he asked again.

"People come and go as we move around," explained Joba. "We try to follow the herds when we can. Right now I'd say that we have around forty. Well, forty Umae and maybe another twenty-five Hek."

The Hek. The stout warriors from atop the cliff face. Dom had seen them moving with the Umae during his recent travels. He thought back to the message he had received in the Chroniclers' Hall.

"Do you have.....family there?" asked Dom. A loose end he would have to deal with, and soon.

"Yes," said Joba. He noted that the two younger boys were beginning to stir. "My mother is there along with my younger sister."

Dom pursed his lips. *Ah yes. Mother.* "I will once more bear Alitus myself. It will be no burden and it will be my way of honoring his courage."

Clagen and Tybor arose from their makeshift beds, rubbing at their eyes. "You still don't have any clothes," noted Clagen.

"I still don't need any," answered Dom.

"What happened to yours?" asked the child. "Did you lose them?"

Dom pushed his irritation down into the well where he restrained his growing fire before it manifested itself in some regrettable way. "You wouldn't understand," he said finally. "And you shouldn't question a Journeyer like that," he scolded. "Eventually, I will be sure that you know everything you need to know."

Tybor frowned. "What's a 'Journeyer'?"

"Come on," said Joba, "leave him alone. We have a lot of walking to do. You can bear him if you so choose, but I can help you if you need it."

"I will simply carry him," said Dom. "Lead the way. Once we arrive at your camp, I have much work to do." Dom crouched down and effortlessly scooped the body up into his arms.

"But I'm hungry!" complained Clagen.

"We will eat soon enough," said Joba. "We need to reach camp before the body attracts danger." He glanced uncertainly at Dom. "Right?"

Dom nodded. "Correct."

Joba, his spear in hand, took the lead with the two boys trailing along close behind. Dom followed at the rear easily bearing Alitus's oversized form. They all moved off in the direction of the hills in the distance.

The grasslands quickly gave way to a nearly grassless plain covered with rocks. Everyone except Dom carefully maneuvered past the rocks to avoid tripping. Dom seemed to take no notice of them at all as he easily moved over the terrain. It was apparent that Clagen and Tybor spoke only Hek, so Dom used only that language. At the far edge of the plain arose a forest of sorts with sparsely scattered trees. Once they entered the forest Joba stopped and knelt down. "Look, here are some more," he said to the boys. The boys anxiously hurried over to look at the ground in front of him. They saw a pattern of wedge-shaped impressions in the soft dirt.

"Was this from the same one?" asked Clagen.

"I doubt it," said Joba. "The other set was pretty far away."

Dom peered down at the pattern in the dirt. "I've seen a number of the creatures who leave these imprints," he explained. "They are quite common."

"Creatures?" Joba flushed slightly. "'Deer'. We call them 'deer'".

"Of course," said Dom flatly. This specific term had not arisen in conversation yet. This was Dom's first opportunity to translate it. "What else would you call them?"

Joba shrugged his shoulders and resumed walking. The trees were far enough apart that they posed no real impediment to progress although they did provide intermittent shade from the sun. This gave Dom no relief. He burned inside as he had every rotation since his mate's death. It was a fire he planned to use.

Just as the boys began to complain and ask for a chance to rest, the group heard voices calling out ahead of them. "That's them!" said Joba excitedly. "Come on." Energized by the discovery, they hurried forward until they ran into a group of six Hek, all bearing long spears.

"Joba!" grunted one with relief. He approached him and took his shoulder in his hand. "You are well."

Joba lowered his eyes as the newcomers noted Dom and his burden.

"What is this?" asked one of the men. "Alitus has been injured and is borne by a child?" The Hek stood looking dumbfoundedly at the naked Umae child carrying their companion. "Alitus is dead," said Joba finally. "Killed by a daggercat."

The Hek began chittering excitedly to one another.

"Are there others?" asked Dom as he set Alitus on the ground.

"Other what?" asked one Hek. He turned to his companions. "He makes our words."

"Of course I do," said Dom. "I am a Journeyer. There is almost no limit to what I can do. And I plan to teach all of you many, many things."

"He frightened the daggercat way," said Joba. "He saved our lives." The Hek exchanged glances but were at a loss to express their confusion.

"Are there other groups out looking for them?" asked Dom.

One of the other Hek nodded. "Yes. Joba, your mother is with another group. They went towards the Lake of Green Waters. She has been gone for nearly 3 rotations."

"Is that what you call the lake that is both heartside and morningside from here?" asked Dom.

"Oh no," protested Joba. "Is there even a lake in that direction? The one we are talking about is heartside, but much more eveningside. It's the one at the far edge of the rocky area we walked though, but in the opposite direction."

"I know it," lied Dom. "I simply did not know your name for it."

"We can carry Alitus the rest of the way." said Joba. "My clan is nearby, but you must be tired."

Dom nodded. "I've another matter that I must attend to, so I must leave you for a short time. It is a matter for a Journeyer."

Once again, the Hek chittered back and forth.

"Where are you going?" asked Clagen eagerly. "Can we go?"

Dom glared at Clagen, instantly blunting his enthusiasm. Clagen bit his lower lip and his eyes watered. He moved next to one of the Hek and tried to bury his head in the man's thigh.

"It is too dangerous," said Dom.

"Will we see you again?" asked Joba. "You said you'd teach us."

"And I will," said Dom. "I will easily be able to find you. Now, I must go. It is a matter of some urgency."

Without another word, Dom dashed off amongst the trees, quickly exceeding the line of sight of those he had left behind. He had all of the information that he needed to erase a minor flaw that had developed in his plan. Now he finally had an opportunity to release some of the energies that raged inside of him.

CHAPTER 5

"How long will I be reviewing all of these tutorials?" asked Lin. She sat frowning in front of a large panel in the Life Lab. Sol had been explaining to her the process he planned to employ in teaching her to become a Life Agent.

"The tutorials are only a part of the program," said Sol. "A more personal level of interaction will be necessary as well. I am typically working on any number of research projects. You will also assist me with them."

"But......I don't know anything," protested Lin. "How can I help?"

"For starters, you can help me with our friend Frell here," said Sol. Upon hearing his name, Frell glanced about before returning Sol's smile.

"Is something wrong?" asked Lin.

"Absolutely not," said Sol. "In fact......." Sol paused as he noticed Frell playing with the container of colored liquid. Now the liquid wasn't changing colors. Instead, different levels of the liquid within the bottle had assumed different colors of the spectrum, giving it the appearance of stripes.

"What is that?" asked Lin. "Did he break it?"

Sol shook his head. "Oh my no," he said slowly. "Far from it. Has he ever spoken to anyone that you know of?"

"No, he hasn't," Lin admitted. "Why?"

"He is holding what's called a 'Chromatic Striation Chamber'," explained Sol. "It's a toy of sorts. I gave it to him mainly to keep him from getting bored."

"He seems to be enjoying it."

"He's more than enjoying it," said Sol. "He's actually mastering it."

Lin reached out towards Frell. "Can I see that?" Frell happily handed her the tube.

"Anyone can take the chamber, shake it up, and make the colors change back and forth," continued Sol. "But the color changes take place in discrete patterns."

"What is the liquid made of?"

"It's a combination of amino acids, sugars, and a number of complex molecules needed for metabolism," said Sol. Noting Lin's confusion he added, "Don't worry. You will understand all of these things eventually. The trick to the toy is in identifying the patterns and then adding energy to the tube in the same ratio as the frequency of the color changes. In other words, the quicker the changes, the harder one needs to shake the tube. Once that happens, the contents of the tube react with precision and grants you a prize."

"A prize? What sort of a prize?"

Sol chuckled. "Oh nothing too grand. It's essentially a glob of proto-plasma similar to what we see in pre-biogenesis. You can stretch it and form it into different shapes. You can eat it. Whatever you want to use it for."

"And Frell did this?"

Sol nodded. "It would have had to have been him. I just gave it to him earlier and he's had it the entire time. It is possible that the tube was jostled or dropped and that it was activated like that. If the colors separate later, we will know for certain. But if we get goo…….."

"Then Frell figured it out," said Lin approvingly.

"Exactly. And the number of Umae his age who can do that is extremely small. It would be an indication of a truly spectacular intellectual potential."

Sol's bright expression darkened slightly. "Lin, Eve and I have discussed what might be the best course for Frell. It may well be a number of ellipses before he is ready for Journeyer training. We will have a number of projects to work on and we likely won't have the time and resources to devote to determining the exact nature of his talents. But have no doubt, he has them."

"So what are you suggesting?"

Sol drew a deep breath. "Repose. He's very young and has his entire life ahead of him. In another fifteen ellipses or so you will have completed your own training. Perhaps then would be a better time for him to garner a few more ellipses before testing again. And you would still be here for him when he wakes up. It would also give us a chance to do a highly detailed analysis of his brain. I have a feeling that we will discover some incredible things about our young friend."

A new weight bore down on Lin's expression. She reached out to Frell and he quickly gave her a long hug. "Frell," she said as she drew back, "did you understand?"

Frell took a long look at the tube in his hand before holding it out to Lin. She received it hesitantly. Frell stepped over to Sol and gave the old man a hug as well. Then he walked over to a nearby repose chamber and stood quietly.

Lin sniffled.

"I'd say he did," said Sol. "Interesting." Lin walked over and stood by the little boy. "Frell," began Sol calmly, "I'm going to open the door just like we did before, and then you will step inside, all right? It will be just like going to sleep again. And when you wake up?" He nodded towards Lin. "Lin will be waiting."

"I will be waiting," affirmed Lin. "Count on it."

Frell nodded and waited for the door. Once it opened, he stepped inside. Sol activated the repose sequence and soon Frell was back in repose.

"I'm going to miss him," said Lin quietly.

"How? He'll be right there!" said Sol with a wink. "I just hope I'm still around in fifteen ellipses. I feel sure that a comprehensive evaluation of that boy will reveal some fascinating details. I'll start the analysis immediately. His contributions are likely to be rather novel, I'd guess. But now, you asked if you can help me in any way. The answer to that is a definite and enthusiastic 'yes'!"

"I still don't see how I can help you with anything now," said Lin gloomily. "Can you explain that?"

"Because you don't know anything," replied Sol. He paused to let his cryptic reply sink in. "What should one do when one doesn't understand something?" He began organizing a number of glass containers on a nearby table.

"I'd ask questions, I guess," said Lin. "If it's all right, I mean."

Sol smiled. "'All right'? Of course it's 'all right'. In fact, it is expected."

"I don't see how that helps you."

"Too often researchers, even old hands like myself, can fall into a rut. They do things the same way they've always done them without considering better options. Having a student can help guard against that because they ask questions about everything."

"Everything?" responded Lin. "I'm supposed to ask about everything?"

Sol laughed. As he did so his laugh slowly evolved into a mild, hoarse cough. "Excuse me," he muttered. Reaching into his pocket, he withdrew a small vial. He removed the cap, took a quick drink, and then returned the vial to his pocket.

"Are you all right?" Lin's voice bore the weight of fresh concern as she watched him drink the liquid from the vial.

"I have a lot of ellipses," he explained. "Sometimes it isn't so easy. But no, not everything. When we first begin work on a project I will explain the nature of our work in great detail. That way, you can draw the greatest benefit from not only your observations, but from the tutorials you will be studying as well. As you progress, my explanations will become more and more general and you will be expected to determine the details on your own. As you do that, you may well determine that some of the approaches I'm taking aren't valid. Or, you might not think so. It will be at this point that you will ask the most questions. You see," he said, patting her paternally on the shoulder, "understanding can result from the elimination of falsehoods as well as it can from the introduction to truths. You simply have to understand why the falsehoods are false."

"It sounds like so much," said Lin dejectedly. "What if I can't do it?"

Sol sat down in the chair next to her. "You can. I've been around you long enough to tell. You have everything you need to become an excellent Life Agent. Your greatest limitation, I'm afraid, will be your teacher. But there isn't anything you can do about that."

Lin blushed. "But…..no. That can't……."

"Sol. A word." It was Eve. She was standing at the entryway to the Life Lab. Without waiting for a reply, she walked in their direction, stopping to sit down in front of the lab's Alpha panel. She closed her eyes and placed her palms on the panel's display. A blue horizontal line rose and fell on the display underneath her hands, illuminating their outlines as they passed up and down beneath them. Sol waited impassively. Lin's eyes darted back and forth between the two Journeyers, eager for someone to explain what was happening. After a few moments, the line stopped moving. Eve opened her eyes and removed her hands from the display.

"What was that?" asked Lin.

Eve frowned at Sol. "It's fine," said the old man. "I told her she was allowed to ask questions."

"Information from the **Aurora**," said Eve. "A lot of it. Jack allowed me access to the logs on the other ship so I searched them and absorbed as much of the data as I could. I just stored the data in the logs of the **Starshine**."

Perhaps too overwhelmed by what she was hearing, Lin remained silent. "What did you want to talk about?" asked Sol.

Eve glanced at Lin. "Lin, I'll need for you to leave for now. This conversation must be private."

Lin quickly stood up, her face turning red. "I'm sorry……." She began.

"Don't be," said Sol. "You haven't done anything wrong. But Eve and I are both officers of the ship, and you are not. Not yet. But before you go……." Sol opened a drawer and withdrew a silver bag and handed it to Lin. "Your Life Agent's bag," he said smiling. "It's traditional for a mentor to gift a Life Agent's bag to his student. May you find many uses for it. Lin glanced inside and saw a number of various devices, none of which she could identify. "Don't worry," added Sol. "You will learn them all."

Lin's face lit up as she looked at her gift. "What should I do with this?" she asked, holding up the vial Frell had given her.

"Your bag can contain whatever you'd like for it to contain," said Sol. Lin carefully placed the vial inside the bag.

"Thank you, so much. I hope I'm half the student you seem to think I'll be."

"I have no doubts," replied the old man. Lin nodded and hurried off to the doorway Eve had entered earlier.

"Sol, Jack offered me access to the logs so I could analyze the data he was talking about earlier. The hidden beacon signal. The information being transported away from Uma. The deflection of the signal by some as yet unknown means. But I stored that data in my subconscious. I wanted to actively analyze Jack's actions generally since we left Uma. I had access to that information as well. I think it is somewhat disturbing, but I wanted to consult with you before forming an opinion."

"So?" asked Sol. "What was he doing?"

Eve sat back in the chair, letting loose a slow breath. "He almost immediately set about upgrading just about every system on the *Aurora*. Based on the schematics he showed us he was incredibly successful."

"And what's wrong with that?"

"Nothing. But it was the way that he did it. He spent the better part of twenty ellipses interacting with the virtual personalities of Min and Hab. He would devise a method to upgrade a particular system all by himself, and then he would ask sim-Min and sim-Hab how they might go about the same task. If they had a better way, he would record it and then erase the portions of the logs describing what Min and Hab told him. Then he would implement the upgrade the way they had told him to before asking them again how they would go about it. Of course, since he had already done what they were going to tell him, they had nothing but praise for him and his ingenuity."

"Unusual," agreed Sol. "But I don't see why this is a reason for concern."

"Sol, he tried to erase those portions of the logs. This had two purposes. For one, sim-Min and sim-Hab wouldn't be able to access the prior discussions and would determine that Jack had come up with the upgrade method all on his own. Second, I think he wanted anyone who ever accessed the logs to think the same thing."

Sol's eyes narrowed. "Why so?"

"The methods he used to erase the logs were very, very complex. Far, far more complex than what would have been required to simply shield the records from sim-Hab and sim-Min. I was only able to find them because I'm a Data Agent and he isn't. He now believes that he did develop those advancements on his own. He believes they weren't actually the result of a cooperative effort between the three of them."

"Oh come now," protested Sol. "How can you know that?"

Eve closed her eyes and collected herself. "Because of the log entries from afterwards. Sol, he went on and on explaining how inferior Min and Hab were to him, and about how his was the keenest technological mind in Umae history. He finally banished the Hab simulation entirely. It isn't even available on the **Aurora** anymore. And the Min simulation….." She paused to swallow. "He transformed it into the artificial intelligence platform for his new mechs. They are all Umaeoid. They all speak with Min's voice. And they are all absolutely subservient to Jack."

Sol was already lost in thought, considering the implications of what he had heard. "So, he spent nearly twenty ellipses working with and, in his mind, exceeding the two greatest technical minds he had ever known about. And this might be critical – he did it without the companionship of any other Umae. Instead of going into repose, he did this instead. And at the end of this period, when he finally did enter repose, the **Aurora** was far advanced from where she had been. His ship. Far advanced, in fact, from the **Starshine**. Min's ship. I'll need to go back and review his behavioral records from when we first encountered him after our repose cycles ended."

"What for? What are you looking for?"

Sol grimaced. "I'd rather not say yet," he said finally. "I'll need some additional data. It could be nothing."

So the training began. Mac went aboard the **Aurora** to become Jack's Technical Agent apprentice. After a great deal of debate, Sol and Lin joined them after Jack was able to convince Eve that the research facilities on the **Aurora** were vastly superior to those on the **Starshine**. Sol himself put up very little argument and almost seemed eager to transfer to the other ship. That left Eve and Lise on the **Starshine** where Eve commenced Lise's studies with an intense series of basic tutorials from the Alpha panel. Lise's baseline of scientific and mathematical knowledge, as determined by her initial examinations, was much lower than Eve had expected. Upon additional reflection, it made more sense to Eve that such was the case given the cultural bias against all of the sciences on Atla. But with her customary resolve, and with a thorough series of data studies to determine the best way to proceed, Eve pushed Lise through her lessons with vigor.

Sol and Lin spent the great majority of their time in the **Aurora**'s Life Lab. While Lin proved to be an outstanding student both in terms of ability and enthusiasm, Sol's instruction methods differed greatly from those Eve employed. Certainly, there were a large number of tutorials for her to endure, but Sol also spent a great deal of time just talking to her. Many of their discussions didn't seem like an education at all.

Lin was most fascinated by Sol's hands-on work in the lab. The tutorials were fine but she wanted to know how to implement some of the things he was teaching her. His primary research focused on what he had described to her as "RTS", or "Repose Termination Syndrome". It frightened her a bit as apparently it was a condition that many Journeyers eventually developed.

"How long have you been working on this?" she asked. Sol was bent over a viewer studying the interactions of a specific group of compounds. He stood up, absently rubbing the white whiskers on his face.

"First, let me explain to you something of the ethical obligations of a Life Agent." He motioned for her to sit, as he often did before settling into a lengthy discussion. Once they each found nearby chairs, he continued. "One of our duties is to attend to the health of our fellow crew members. They must feel free to share with us everything they can about their health. Otherwise, our objective data might be compromised, which in turn compromises our effectiveness in treating them. How do you suppose we do that?"

Lin enjoyed these puzzles. Rather than being spoon-fed, he led her halfway to an answer and then forced her to find the rest of it for herself. "Well, I still don't know what technology you have available. Is there some sort of..... truth machine?"

Sol broke out laughing. "No, no.... I'm afraid we don't have anything like that. It's much simpler. Try again."

Lin thought hard, wanting to impress her teacher. "I'm trying to think of why they might not be honest if they want you to help them. I don't know....... Maybe they are embarrassed?"

Sol nodded. "Yes. Or afraid. Or both. You are on the right track. So, how do we deal with that?"

"If I were sick, I wouldn't hesitate to tell you everything. You don't make me nervous at all. Just the opposite."

Sol clapped his hands. "Good! I'm glad. And that's part of it. 'Bedside manner' you might call it. Make people feel comfortable and they will open up. But what if they are embarrassed about something?"

"If I were embarrassed about something, I wouldn't want anyone to know," said Lin. "The fewer people the better."

"Exactly right," replied Sol. "But your Life Agent has to know. So......?"

"So......... the Life Agent promises not to tell?" Lin blushed slightly. She wasn't entirely satisfied with her answer but it was the best she had.

"In a sense. The Life Agent has an obligation to the patient not to share any information about a patient with anyone else, with minimal exceptions."

"It just comes down to trust?" asked Lin.

Sol nodded. "And trust, in part, is somewhat related to that bedside manner I described, too. But what if someone has a condition that is very serious? So serious that it could be a threat to the other crew members or to the mission. What should happen then?"

Lin squirmed under the weight of this new question. She didn't see any apparent correct answer. "The crew has to be protected, doesn't it?" she asked meekly. "And the mission? Isn't that why we are out here?"

All hints of levity vanished from Sol's face. "But how does one make that judgment? How can one be sure?" he asked quietly.

His young student stared over at the Alpha panel, anxiously waiting for some intelligent thought to enter her brain. "I don't know. It doesn't seem like one can be sure about something like that," she said finally.

The Life Agent's expression brightened slightly. "Certainty," he said quietly, "is a wonderful, but rare state of mind for this position. Very, very often a Life Agent can't simply rely on test results and readings of various sorts. When you are faced with a decision about a patient requiring you to weigh the patient's privacy against the welfare of everyone else, you will never be certain. At least I have never been. But you will have a sense of what you must do. Remember those aptitude tests you and your fellows took just after we left Uma?"

Lin nodded. "Yes," she said with exasperation. "They took forever."

"I could have told you before those tests were administered that you would be highly suited for training as a Life Agent. Not because I appreciated how smart you are. But because I could sense your empathy for others. You will learn to use that empathy not just to connect with your patients....... bedside manner and all that....... but to understand them. To be able to, if necessary, judge them. There are no tests for that type of understanding. You have it and hopefully you will find someone to help you develop it and understand it."

"Have you had to do that before? I mean, make that type of decision?"

Sol leaned back in his chair. "Lesson two. Can you share patient information for academic purposes? The answer is, 'yes'. Min is still on Uma's moon. He has a very advanced case of RTS. This research of mine, which I've been doing for many, many ellipses, is an attempt to cure that condition. Or at least mitigate its impact. So far, I haven't been able to do it. But.....," he added with a renewed warmth, "I have a great motivation to find a breakthrough. Once we have found Haven, we will return to Uma and can see Min. I hope to have something for him by then."

"You are friends?" asked Lin hesitantly.

"More than that," answered Sol quietly. "But here." He stood up and entered a series of directives into a nearby panel. "During your next tutorial, I want you to review this case study in detail. Let me know when you are finished and we will go over it."

"Are you close?" asked Lin. "To a cure, I mean?"

Sol thought for a moment before a series of coughs began wracking his body. Lin stood up to assist him once she became concerned that he wasn't going to be able to stop. He braced his arm on the table in front of him and drew something from his pocket and put it into his mouth. His coughing began to ease. "Not close enough," he said finally. "But remember, Life Agents have to keep secrets sometimes."

Lise sat in front of a side panel, eagerly waiting for the static pattern on its face to disperse. "Mac? Are you there? Can you hear me?"

The continued static from the panel fueled her frustrations. She fidgeted in her seat, vigilantly searching for any break in the interference. Just as she decided to rise and stomp off, the pattern broke and she could see Mac's face.

"Yes," he said quietly. "But it took me a while to figure out how to use this screen," he said blushing. "Jack wouldn't help me."

"Why not?"

Mac shrugged. "I'm not sure. He said something about how I have to be able to use my brain to solve problems for myself. But he hasn't really been spending that much time with me. He has me watch tutorials while he stays in his room. I don't know what he's doing in there."

Lise sighed. "I don't care what he's doing," she gushed. "I'm just so glad to be able to see you. I miss you so much."

Mac seemed at a loss for words. He smiled uncomfortably and looked away from the screen. "Yeah........"

There was a clatter from somewhere behind Mac.

"What are you doing?" It was Jack and his voice was filled with rage. "I told you not to use that without my approval!" His face appeared on Lise's screen.

"I was the one who started the conversation," said Lise. "Don't be mad......"

"This is my ship!" screeched Jack. "And all communications to and from my ship are subject to my authority!"

"But......" Lise's plea was cut off as her screen went blank. She tried to re-establish the connection but there was no response from the **Aurora**. "Ahhhh!" She covered her face with her hands and set her elbows on the panel. "Mac......."

"What's wrong?" asked Eve. She had just entered the chamber and was still standing in the doorway. "Are you ok?"

Lise was trying not to cry. "I guess. I was talking to Mac. Jack got angry and cut off our conversation." She slowly turned towards Eve, uncovering her face to reveal tear-streaked cheeks.

Eve took a slow breath and then approached Lise. "You might not believe me but I do understand how hard this is."

"I don't believe you."

Lise still didn't have very many ellipses but had already learned something about that special connection that one Umae could develop for another. Eve hadn't had any such experience before she met Min. But now Eve knew a great deal about the consequences brought on when that connection was lost.

Eve placed a gentle hand on Lise's shoulder. "Believe me," she said quietly. "Our obligations as Journeyers sometimes force us to make hard choices. Choices that we might wish we didn't have to make. But once you consider how important our mission is......."

Lise looked up at her mentor. "I'm not that important, am I?'

Eve frowned. "Don't think of it like that," she said quietly. "We are all important. But we are also the only ones who can save our people. That is more important. It has to be."

"Why?" Lise's simple question twisted its way into Eve's gut. "What happens if we fail? There aren't any more Umae? Why is that so bad? We give up everything to save the lives of people we will never know. And for what? We get some sense of satisfaction from a job well done before we die and disappear into oblivion? And if we do fail, we have given up everything for nobody. There won't be one single Umae to care. I know I haven't been your student for very long, but that seems rather stupid to me."

Eve searched for a suitable response. She could only think of the monitor that had relayed Min's vital signs to the **Starshine**. The monitor that had reported the news of his death to her. He had encouraged her to go and to do her duty. He told her that there was still a chance to both save the Umae and to be together. A chance for everything to be right. Now he was gone and she found herself caring less and less about the mission. But the mission was all that she had left. If the mission failed, Min's death would end up being worthless. She wouldn't let that happen.

"There's still a chance for you and Mac," she said finally. "To make everything right. You can save our people and you can still be together. Do you think you would be happy just being with him if you knew the Umae ceased to exist because of it? Could you live with that?" Eve was asking Lise many of the same questions she had long been asking herself.

Lise lowered her eyes. "I guess not. But…..how much longer? Before we get to where we are going, I mean?"

"We don't know for sure," admitted Eve. "The scans from the *Aurora* indicate that the deflection of the tether beam occurs about thirteen light ellipses from here. Once we find out how that is happening, we will follow the tether to wherever it leads us."

"But can't the *Aurora* just tell where the tether goes after it is deflected? Then we could go straight there?"

Eve smiled. "An excellent point. I'm impressed with your logic. But unfortunately we can't do that. The angle of deflection is away from our course. The tether eventually exceeds the range of the Aurora's scanners. And we don't know if there are any other deflections we haven't detected yet." She thought of all the technical upgrades Jack had made to the *Aurora*. "At least we don't right now. Maybe once we get closer to the deflection point, or maybe if Jack works some more on the *Aurora*'s sensor arrays, we can find out. But for now, we follow the tether."

"And I get to see Mac on the display," added Lise. "That's not fair."

"No," agreed Eve as she gave Lise's shoulder a gentle squeeze. "It's not fair. But that is a luxury we don't have right now. Maybe someday." Again, the memory of the silent relay monitor intruded into her mind. "But maybe not. Until then, we do what we can do."

Mac's face was twisted with anxiety. He was working on a panel tutorial while Jack peered over his shoulder at the display. It seemed that nearly every entry the trainee made resulted in a sharp retribution from his teacher.

"Come on, Mac!" barked Jack. "Did you actually take that last tutorial? Your assessment indicated that you are at least of above-average intelligence. Did you cheat? Were you just having a good rotation when you took the test?" Jack jerked the arm of Mac's chair rotating so he could stare his student directly in the face. "You need to improve, and quickly," snapped Jack. "I've got other things I need to be working on and you are rapidly becoming a waste of my time. When you turn out to be an abject failure as a Technical Agent, it's not going to be MY fault!"

Mac lowered his eyes, wanting nothing more than for Jack to just stalk off as he typically did after one of these tutorials. "I.......didn't understand everything in the tutorial," he mumbled.

"Well, that's obvious!" screeched Jack. "A truly great understatement." He took a deep breath. "Maybe you just aren't cut out for this," he said in a calmer voice. "Maybe we can find something else for you to do."

Mac looked up. "Like what?"

"Your technical aptitude is your strength, unfortunately. The assessment demonstrated that you would be completely lost in life science training or data studies. Since that is really all that we need on a Journeycraft, maybe we could test you to see if you are qualified for.... I don't know. The mechs already take care of all the cleaning, so that's out. Do you think you can hold things? Like scanners, or spare items of clothing? Because I think that would pretty well maximize your intellectual capacity!"

Mac lowered his eyes again, waiting to see if Jack would just leave.

"Jack?" It was Sol. He was standing at a nearby door with Lin. "May I have a word?"

Jack stood up, straightened his shirt, and approached the Life Agent and his student. "Do you need some extra mass in the Life Lab?" he asked. "In case a wind suddenly kicks up? Some of your lighter instruments might be blown about. I can recommend Mac for that position. I feel confident that he could be fairly proficient at weighing things down."

Lin looked away, uncertain about her ability to mask her disdain for Jack's behavior. "Training is always difficult," noted Sol in his usual gentle tone.

"It wasn't for me," countered Jack. "In fact, Min had a hard time finding tutorials that challenged me. Mac has difficulties turning the panel on some rotations."

"Well, keep on encouraging him," offered Sol. "He'll catch on."

"What do you need?" asked Jack impatiently. "I have things I need to be working on."

Lin turned back towards Jack, her face a model of perfect composure.

"We are doing some rather complex research," explained Sol.

"Still on the RTS treatments?" asked Jack.

Sol nodded. "Yes. The data I gathered from Hab has provided me with some additional insights which I hope will be productive."

"But?" asked Jack.

"I was technically never named as the Life Agent for the *Aurora*. Without that designation, I can't access some of the more powerful analytical platforms offered by the Life Lab's panel. It's starting to impair our research."

"'Our'?" asked Jack as he looked at Lin. "Is she actually helpful?" Skepticism dripped from his tone.

"Absolutely," replied Sol firmly. "I've integrated her tutelage with work on the RTS research. She's learning theory and practice at the same time." He nodded in Mac's direction. "Maybe......."

Jack raised his hand towards Sol. "Stop! Don't tell me how to instruct my student. Apparently, you were fortunate enough to get one who doesn't have a severely limited intellectual capacity. I'll teach him using my own methods. I'm not optimistic though. But if he doesn't start making better progress, I think my time will be better spent on other projects."

"'Projects'"? protested Lin. He's not a 'project'. He's an Umae. I've known him his entire life. He may not talk very much, but he is quite smart. Trust me."

"In your expert opinion?" mocked Jack. "Did you assess him? Did you evaluate his scores? Maybe you should just stick to stirring beakers for Sol. You'd do well to remember the hierarchy of command on this ship. If I want your opinion about something, I'll ask you. But don't hold your breath."

Sol coughed weakly, enough to rid himself of a tickle in his throat. "Can you appoint me Life Agent?"

Jack smiled. "Yes, of course. I truly hope you are successful in your research. In fact, if I can find time," he added, glancing momentarily back at Mac, "maybe I can help. I'm sure that would greatly increase your odds of success. I'd love to be able to cure Min. There are a lot of theories I'd like to explain to him. Some of the things he believed......?" He wrapped his arms around himself and simulated a shivering man.

"I remain optimistic," said Sol, the offending tickle now apparently banished. "The sooner the better."

"I'll do it immediately," said Jack as he headed off towards the Alpha panel.

Sol turned and started the walk back to the Life Lab with Lin following behind. As she began to speak, Sol raised one finger bidding her to remain silent. Once they entered the Life Lab and the doors sealed behind them, he turned and faced her.

"You have questions."

Lin nodded. "Has Jack always been like this? I mean, the way he has been since we came out of repose?"

Sol slowly shook his head. "No, he hasn't. Jack has always been brilliant and enthusiastic, but he has also always been respectful and considerate of others. Those last two qualities seem to be gone. He has also always been exceedingly confident in his talents, and justifiably so. Min always marveled at Jack's intellectual prowess. But it seems Jack's confidence may have risen to dangerous levels."

"What do you mean?"

Sol pointed to the Life Lab's Alpha panel. "I gave you a case file for a former crew member of the *Aurora*. His name was Anth. Have you reviewed it?"

"Briefly. Should I go back and review it again?"

"Yes. And do so as thoroughly as you can. There may be terms or concepts in there that we have not studied yet. If there are, use the panel to help instruct you on those matters. When you are finished, find me. I wouldn't be surprised if your review took you 7 or 8 rotations, so don't worry about hurrying."

"Do you have some new tests you are going to run now?" asked Lin.

Sol shook his head. "No, why?"

"You told Jack that you needed access to new platforms in the panel. Wasn't that why you wanted to be designated Life Agent for the *Aurora*?"

Sol chuckled. "You are observant! I do need to be designated as Life Agent, but not just because I want to access the platforms. Go and finish your research and we will talk some more. And be mindful of a Life Agent's charge."

"The well-being of the crew?"

"Yes, the well-being of the crew. But beyond that?"

Lin frowned. "The mission?" she asked finally.

"And what is the mission?"

"We are looking for Haven. We are......."

"Why?" asked Sol with a sudden burst of vigor. "Why do we seek Haven?"

Lin could sense the importance of this lesson and desperately wanted to give Sol the right answer. "To save our people."

Sol put an arm around her shoulder and nodded his head. "Yes, yes. We Life Agents, first and foremost, must act to serve our people. All of them. Don't ever forget that."

CHAPTER 6

The training continued on both ships. Sol combined the use of the panel tutorials, work in the lab, and frequent one-to-one conversations with Lin. Jack mostly ignored Mac, spending most of his time locked away in a room working on a project he wouldn't talk about. On the **Starshine**, Eve constantly had to redirect Lise to her studies. Her mind tended to drift off towards the **Aurora** and her next comm-chat with Mac. To some degree, all of the students had made progress in the ten ellipses since they had awakened from repose sleep. Mac struggled the most for essential lack of an instructor. He repeated the panel tutorials over and over, trying to fathom the mysteries of Journeyer technology. Any interactions he had with Jack were critical. Mac was stupid. Jack couldn't understand how Mac was the Technical Agent trainee. Mac would never be useful for anything other than basic data entry. All of these, and many others, were Jack's opinions that he rained down on his beleaguered student whenever possible. To a certain extent, Mac had grown resistant to such treatment. His talks with Lise helped him a great deal. Mac was a quiet young man, but Lise's constant attentions were slowly drawing him out of his shell. He was persistent and looked very much forward to seeing her in person again.

When Lise could focus, Eve was impressed with her progress. But focus was a critical skill for a Data Agent. They had not yet attempted to interact with the panel via tactile interface and Eve remained concerned that Lise's wandering mind might not be up to the task. Eve attributed that, for now, to Lise's youth. Eve's youth had been rather different. She had been much younger than Lise when she first boarded the **Starshine**. There weren't any boys her age, and she was not yet old enough to see them as anything special. But as she grew, she spent more and more time with Min and all of that changed. By then she had finished her training as the ship's Data Agent and had assumed that role. She suspected that if she had been experiencing the growth of the relationship she'd had with Min while she was trying to train, she would have been distracted as well.

Lin was a star. Her advancement was far beyond anything Sol could have hoped for. Not only could she handle the theoretical side of her training, she was adept in the lab as well. But most importantly, at least for Sol, was her empathic sense for others. Although she didn't say much about it, Jack's treatment of Mac bothered her. Mac was resilient, but he frequently became depressed following one of Jack's senseless tirades. Sol didn't intervene. Lin wondered if that was a part of her training or if it was something else entirely. While Sol seemed to be a simple, gentle old man, his actions were measured. Everything he did had a reason behind it. Part of Lin's training had been to keep up with him, trying to understand what he was up to.

A detailed evaluation of Frell's neural functions had been implemented as a part of Lin's tutelage on that topic. Sol was stunned by what they had discovered. Frell's neural connections were extremely efficient in both design and function. In fact, Sol was hard pressed to determine how he might have better designed them if he had begun from scratch.

"So why doesn't he talk?" asked Lin. "Why didn't he do better on the evaluations?"

"Frell was raised as a typical Umae child," Sol explained. "I'm sure that his Rearers did their very best to encourage him and stimulate his mind. But his mind is anything but typical. He would have required a very different approach to maximize his development. His Rearers, of course, couldn't have known that and wouldn't have been able to do it even if they had."

"Is it too late?" Lin looked fondly at the boy in the repose tube, a hint of concern in her voice.

Sol shook his head. "Not at all. In fact, I can develop some protocols for you to implement after he awakens that should result in a normalization of his behaviors without impeding his intellect."

"Me?" asked Lin.

Sol smiled. "Of course, you. He trusts you. He already looks upon you as his Rearer. And you will be uniquely qualified to do it." He placed a hand on Lin's shoulder. "Imagine what sort of an impact he could have on the Umae. All societies develop at different rates, but all of them also benefit from periodic bursts of ingenuity provided by truly gifted individuals." He looked over at the tube. "I'm confident Frell will be such an individual. With your guidance, he will be a gift to the Umae people."

"I might need you to help me," said Lin uncertainly.

Sol paused. "Of course. As much as I am able. But I'm not getting any younger, you know."

Late in a rotation, some thirteen ellipses after training had commenced, Jack emerged from his lab. His face practically glowed.

"I've got it," he announced, waiting for everyone's attention. Mac looked up from his tutorial and glanced over at Sol and Lin. The Life Agent and his student were using the ship's Alpha panel to analyze data from their most recent trial on their RTS research.

"You have what?" asked Lin finally.

Jack walked past Mac's seat, grabbing the young man by his collar. Mac stumbled up from his seat, practically falling to the floor as he tried to keep up. Jack pointed at the Alpha panel's display. "I need this," he said as he halted Sol's analysis. "Watch what I'm doing!" he snapped at Mac. Mac peered sheepishly over his shoulder.

Jack quickly entered some commands into the panel and the display came to life. It showed a variety of bright, angled lines rotating three-dimensionally. "Mac, you lout, I'm sure you don't even know......."

"It's a vector chart," said Mac. He looked a little closer at the display. "The *Aurora* is there and the *Starshine* is over there," he added, noting a couple of blips on the display.

"I'm a better teacher than I thought," said Jack. "But let me tell you what you won't be able to figure out for yourself." He pointed to a pair of broken lines that met at a sharp angle to one another. "This vector is the one the ships are following. But this vector," he continued with a gushing satisfaction, "this vector is the direction of the signal coming from the beacon satellite that is at the juncture of the two vectors. This vector," he stated, pausing for effect, "will lead us to Haven. I'm sure of it."

The others silently leaned in to study the chart. "You are certain?" asked Sol. "No doubt?"

"I know what certainty is, old man," said Jack. "Yes. Absolutely. It is the only possibility. I've checked it and rechecked it."

"We need to tell Eve!" said Lin.

Jack nodded. "Yes. Yes, I do."

Jack activated the comm device he had installed on the *Aurora*. It immediately hailed the *Starshine* and Eve's face soon appeared on the display.

"Time for Mac and Lise again?" she asked.

"No. It's time for us," said Jack. "I've been able to track the deflected signal as far as I need to. There are no other deflections. I've found the path to Haven. We will untether the *Aurora* from this signal and join the second signal on a hypotenuse that I've already calculated. Once we reach the second signal, we will follow it until we reach Haven. You will take the *Starshine* to the beacon satellite."

"Wait!" said Eve. An ember of anger heated her voice. "We have to talk about this. You sound like you have already decided on a course of action."

"I have," said Jack matter-of-factly. "We need to reach Haven as quickly as possible to determine if it is suitable for environmental modification. It makes sense to take the **Aurora** because it is so much faster than the **Starshine**."

"Why don't we both go there then?" asked Eve. "We will need both ships. One to begin the modification process and the other to return to Uma."

"No," said Jack. "That satellite has been relaying data from Uma probably since the Cataclysm. If it works like our satellites do, and I'm confident it does, you should be able to access all of that data. Every single bit of data concerning everything that has happened on Uma since data storage has been an option." He leaned in slightly, grinning madly at Eve. "And I'm guessing that you want to do that. I'm right aren't I?"

Eve's chin quivered slightly. All those answers… "But how long will that take?" asked Eve. "To record the data and then join you on Haven?"

"You might not be able to," conceded Jack. "I'm not sure how far Haven is from the satellite. I only know the direction. If we have determined that Haven is modifiable, it would be most efficient if we simply returned to Uma immediately. It would be up to you and the **Starshine** to effect the modifications."

"You can't do that!" Lise appeared behind Eve, her face glowing red. "I'd never see Mac again. And Eve! What about Min?"

Jack's countenance did not falter. "Saving the Umae is more important. Eve knows that. And you should know that if her tutelage has sunk in. We don't have time for those types of considerations."

"Jack." It was Sol. He put a light hand on the Command Agent's shoulder. "You don't know how much time we have. And you don't know if Haven will be modifiable. Aren't you assuming a lot?"

"Aren't we all?" said Jack as he rose from his seat. "We are out here, almost five hundred light ellipses from our home, because some ancient set of laws tells us that the salvation of our people is nearby. Does it make sense to come all this way only to diverge from the path that Journeyers have followed for……. Who-knows-how-many ellipses? Does it!?!"

The others were silent. Eve finally spoke. "No. It doesn't," she said quietly. "All Journeyers have put their personal interests below the interests of the Umae generally. We should be no different." She drew a deep breath and placed an arm around Lise's shoulder. "I'm sorry, Lise. This is the hardest lesson I'll ever have for you."

Lise twisted away and ran off, her sobbing cries audible on both ships. "I have already made the necessary calculations you will need for the most efficient data transfer," said Jack. "I've also plotted a projected course for you in case you lose the tether beam for some reason. I'm transferring those to you now."

"I'm sure you are," said Eve in a low voice.

"I trust that the *Starshine* has directive strategies for bio modification at that level?"

"They do," interjected Sol. "Helped write them myself." He leaned back and mostly stifled a cough from his chest.

"Very well," said Jack eagerly. "Eve, the Umae are in good hands. I hope I get to see that data someday."

Before Eve could respond, Jack cut the connection and the panel went black.

The *Starshine* had been abnormally quiet. Without any comm-exchanges with the *Aurora*, and Lise's perpetually gloomy mood rendering her effectively mute, the silence on the ship was broken only by the operation of the machinery and Eve's rare success in getting Lise to speak with her. Eve had also become somewhat withdrawn as she was focusing her energies on preparing to analyze what she suspected was an immense amount of data within the beacon satellite. They would be within range soon and she would need Lise's assistance. It was going to be a one-time opportunity for both of them to interact with the single greatest data storage mechanism the Umae had ever encountered. Eve thought she was as ready as she could be. Lise, unfortunately, was not.

Lise spent her time mindlessly staring at tutorials or sitting alone in her chamber doing Eve knew not what. The Command Agent had chosen to respect Lise's privacy. But that privacy was endangering a very important operation. Eve couldn't afford to allow that endangerment to continue.

The door to Lise's chambers slid open, causing her to start. Lise glared at Eve as she entered the room. Lise then placed her head back onto her pillow.

"What do you want?" she groused. "And how did you get in here?"

Eve treaded gently. "I'm your Command Agent. I can go anywhere I want to. We need to talk." Eve moved to a chair near the bed and sat down. Lise pretended not to notice. "We are going to be within range of the beacon satellite soon," continued Eve. "We will have an enormous amount of data to sort through and order. I'm going to need your help."

The air was thick and tense. Eve waited patiently for a reply but wasn't sure how long to wait. Fortunately, Lise broke the silence. "Why?"

Eve exhaled gently. "Because if I do it all by myself, it will take far too long."

"No. I didn't mean that. Why are we doing it at all? Just because Jack said so?"

Eve had to admit to herself that Jack had been convincing. Min had had so many doubts about the Protocols and the Directors for so long, but he never had much of an opportunity to study them in a meaningful way. If Jack was right about this beacon satellite, they would have all the answers to every question they had ever asked. Although Lise likely wouldn't understand, that meant a lot. A lot not just to Eve, but to the Umae.

"We have long gaps in our history, stretches of time when we don't know what happened on our planet. Or what happened to our people. Our first priority is to serve our people. If we are going to successfully find a new place for them to live, we will need to fill in those gaps. Our history is part of a foundation upon which we will build everything else."

Lise slowly sat up and brushed the hair away from her face. "When I was chosen to become a trainee, I didn't really care about leaving Uma," she admitted. "But only because I thought I was going to be with Mac. Now I don't have either, and I probably never will." Eve was surprised at how well Lise was holding up while speaking of this.

"I don't have either....either," said Eve, smiling weakly at her inadvertent play on words. "I thought I was going to be able to stay with Min on the moon. But that didn't happen. So, here we are."

Lise looked up at Eve's pale pink eyes. "It hasn't gotten any easier?" asked Lise. "That's.....kind of scary."

Eve put her arms around Lise and the two women embraced. "Not easier," said Eve in between sobs, "different. But, it's not all bad." Eve let go and leaned back. "Thinking of Min makes me smile. Much more than it makes me sad. It's worth it. I'd do it all again, without question." Her momentary smile faded. "But you and Mac aren't done yet, not like we are. If we see them at Haven, or if we see them back on Uma......."

Lise brightened slightly. "And Min is on the moon, right? Isn't Sol trying to come up with something?" Eve shuddered slightly.

"Yes," she muttered finally. "He's on the moon." She tried to force the image of Min's lifeless body hanging upright in a repose tube out of her mind.

"We have hope then," said Lise with heightened enthusiasm. "Both of us. I'll try not to give up if you don't. Deal?"

Eve swallowed hard and nodded. "Deal."

"All right," said Lise with a hint of renewed energy. "Show me what you need for me to do. Let's get this done and get to Haven as soon as possible. And then? Then we can go back to the moon."

Eve tried to fake a smile. "We'll see. Ok, let me show you something." She rose and started for the door, expecting Lise to follow. Eve led her back to the main bridge area and sat down in front of the Alpha panel. She entered a couple of short directives and an image arose on the panel. "Jack said that he was able to complete a few very basic scans of this beacon satellite. The data we have on it is rather minimal, but one thing is certain. It is huge."

The image on the display was of a long, cylindrically shaped structure with a large concave cup at either end. "How big is it?" asked Lise.

"A typical beacon satellite is about half the size of the **Starshine**," said Eve. "This one is at least five or six times as large as the **Starshine**."

Lise peered at the image. "Why would it need to be so big?"

Eve leaned back in her seat. "It could be for any number of reasons. Depending on the level of technology it employs, it might need to be that large to process and store all of the data it is relaying. If there are only a few beacons relaying to Haven, its size might be a protective aspect of its design. The bigger it is, the more durable it would be in terms of resisting impacts with debris, radiation bursts, and the like. It doesn't move. So it has to be able to tolerate whatever comes its way."

"But, don't you think the Directors built it?" asked Lise. "Wouldn't its technology be highly advanced?"

"That seems like a reasonable conclusion," agreed Eve. "But we don't really have any idea about what type of technology the Directors have. We assume that, at least in general principles, it is similar to ours but we don't know where they are in terms of advancement. For all we know, we have surpassed them."

Lise's jaw dropped. "We.....surpassed them? But that's impossible. They are the Directors, may they guide our way."

"Until Min voiced his doubts about the Directors and the Protocols, I would have had the same reaction," said Eve. "But we will have a better idea once we get to this satellite."

"How far away are we?"

"Just under 5 rotations. The **Starshine**'s data processing capabilities are first-rate, at least in comparison to any other ship I've ever heard of. Min and I spent a lot of time optimizing those systems. If we are close to, of if we have exceeded, the Directors' technology levels we will be able to process the data in that satellite pretty quickly. If we aren't, well, we will be there for a while."

Lise frowned at this possibility. "Doing what?"

"There may be security safeguards to breach, the information itself might be encrypted or corrupt, and those are just the foreseeable issues. Any of those barriers will slow us down a great deal. If we have multiple issues......"

"But I want to get to Haven," said Lise firmly. "I'm going to get there and see Mac."

"All the more motivation for you to help," said Eve.

"Tell me how."

"Are you familiar with our uplink processes? If not, you'll need to familiarize yourself as soon as you can. We will be analyzing, sorting, and interpreting as much data as fast as we possibly can. We will have to derive algorithms as we go to figure out what is junk and what is important. I can do all of those things by myself, but it will go much more quickly if you can help."

"I can," said Lise. She had declared her competency without hesitation. "Just tell me what to do."

Eve put a hand on her knee. "For starters, you need to focus. I know how much you want to see Mac. Believe me, I know. But those types of thoughts can be intrusive. I may need for you to do some work on the neuro-tactile interface. If that's the case, you can't be mooning over Mac. As difficult as this sounds, the less you think about him, the more likely you will get to Haven in time to see him."

"You're right. That does sound difficult."

Eve entered a few new directives into the panel and the image of the satellite disappeared. "I have set up a few basic tutorials about the n-t interface. The concept is rather simple, the execution is not. You have to open yourself up to the possibility of making your mind a part of the device that you are studying. All of those focus disciplines that you have been studying will come into play here. Our minds can do things our machines can't, but only with focus. I wasn't planning on having you try this yet, but we don't have any choice."

"Is it dangerous?"

"It can be," confirmed Eve. "But only if you try to access a measure of data that your mind isn't ready to process. There are numerous safeguards on the tutorials I'm giving you. You will be completely safe. They will also give me an idea about how much you can handle."

"I can handle a lot," said Lise confidently. "Just wait and see."

Eve rose. "I hope you are right. Get to work. I have other preparations that I need to attend. I'll check back with you later." Eve walked off in the direction of an exit at the far side of the bridge. Once she was gone, Lise took her seat in front of the panel.

"So, what now?"

"Place your palm on the display." It was the voice of the Eve simulation. "Then concentrate on whatever information the panel is sending you. And remember. Focus is the key."

Lise nodded to herself. Focus. Goodbye for now, Mac, she thought as she reached for the panel. But I'm going to get this right for us.

Eve and Lise sat next to each other at the Alpha panel, both of them obviously anxious. "Ready?" asked Eve. Lise nodded, almost imperceptibly. "Panel, view target on display," said Eve.

The display came alive, opening a virtual window for them into space. In the center of the display was an oblong structure with a rounded concave curvature at either end.

"Is that what they look like?" asked Lise.

Eve nodded. "Yes, for the most part, except.......one moment." She closed her eyes and placed her palm on the display. Lise watched Eve closely, noting the fierce concentration on her mentor's face. After a short time, Eve opened her eyes and removed her hand. "It's.....enormous," she said, almost breathless. "Its size could be an indication that its technology isn't as advanced as ours. It's possible that this was as small as its builders could make it and still have it be able to do its job properly."

Lise looked back at the image on the screen. "What IS its job?"

"We will need to do some analysis. If Jack's theory is correct, this beacon satellite is designed to relay every tether beam it encounters to Haven. So that anyone who is on Haven now will have a complete understanding about what has been happening on Uma."

"There are other tether beams?" asked Lise.

"Yes, there are several that emanate from Uma. But, it's curious. The others all go off in different directions than the one we use. So something must be deflecting them in the direction of this satellite."

"There are people on Haven already?" asked Lise. "Now I'm confused. I thought we were trying to find it so we could move all of the Umae there. Who would be there now?"

Eve turned towards Lise. "The Directors might be there. The authors of the Protocols. The providers of our base technologies. The creators of the plan to save our people. They might be there. It would make sense. I hope they are."

Lise was somewhat ashamed to admit her confusion to her teacher, but her curiosity won out. "I don't understand. If they were going to help us get away from the Hek and move us to a new planet, then why hasn't anyone ever seen them?"

Eve slowly shook her head. "One of the very many good questions we have all asked before," she conceded. "We don't know who, if anyone, has seen them. Only the Chroniclers know that and they have never told anyone. At least not that we know of." Lise began to reaffirm her confusion, but Eve put her hand on Lise's arm, cutting her off. "There are many things we don't know. But the Directors, may they guide our way......." She blushed slightly at omitting that phrase previously, "...have led us to this point, and Critical Closure has likely not yet been reached. Now is not the time to doubt that guidance. We have some work to do."

Lise still had a dozen questions she wanted to ask, but now understood that Eve didn't have the answers she wanted. "What do you want me to do?"

"I need for you to perform some data scans of that satellite. You will need to access a virtual profile of one of our Technical Agents. Have that profile determine the methodology for the scans and then have it compile a profile of the technological aspects of the satellite."

"Does the use of the virtual profiles in reality match the description of their use in the tutorials

"Yes, perfectly."

"Should I just use Jack's profile? He's the only Technical Agent I've ever known."

Eve closed her eyes for a moment. "I'm not sure if Jack's recent behavioral changes will impact the effectiveness of his profile or not," admitted Eve.

"Changes?"

"Jack hasn't always been the way he is now," said Eve. "Before he was curious and enthusiastic, but he was also respectful and open to the possibility that he had limitations. I always thought that openness was a part of what made him such a talented student. But now? Now he is still very enthusiastic, but his time alone on the **Aurora** changed him. I'm not sure what he did all of that time but now he is confident to a fault, I'm afraid. I don't think he even entertains that possibility that he could be wrong about anything. That could be dangerous."

"So what about Min?"

The sound of his name quickened Eve's pulse, but she bore it better than she would have expected. "That would be a great choice," she said finally.

"What will you be doing?" asked Lise as she rose to move to another station.

"I'm going to go over all of the data management protocols in the Alpha panel one more time. Again, if Jack's theory is correct, then every bit of information transmitted by the beacons orbiting Uma has passed through this beacon. If this beacon operates in the same way as the others do, all of that data is still stored inside of it. Gathering it and organizing it will be a massive undertaking."

Lise folded her arms. "You know, I'm a lot more familiar with the data management protocols than I am with the use of the virtual personnel profiles. Why don't you let me have your job instead?" Eve stood up and they locked eyes. Eve's were teary and her face was slightly flushed. "Oh. Ok."

"Yes," said Eve, putting an arm around Lise's shoulder. "That focus, remember? One of the ways that you can prevent Mac from being a distraction is by avoiding anything that reminds you of him. Just consider that to be another lesson your mentor has to offer."

Lise put both of her arms around Eve and the two women embraced. "Not an easy lesson, is it?"

"No, indeed not," said Eve. "But not all of them are." She stepped back. "I'll review the data protocols from another display. Can you do me a favor?"

"Sure," said Lise. "Anything."

"When you are working with that virtual profile, please ask Min to keep his voice down."

Lise smiled. "Absolutely."

Eve returned to the Alpha panel where Lise had been working. Lise had completed the data scans Eve had requested. Fortunately for Eve the architect of those scans, the virtual simulation of Min, had been retired and was completely silent now. Lise was intently trying to understand the lines of data that were streaming across the display.

"I had to make a few changes to the data analysis protocols," said Eve. "But I think they are ready. How did the scans go?"

Lise gestured at the panel. "I'm not sure. Min, well virtual Min, was able to design data scans that have given us a general outline of how the data inside the satellite is stored. You had suggested that the amount of data might be massive. I know that isn't a very objective term, but we may need a new one. What is beyond 'massive'?"

Eve leaned in to view the panel. "Moons," she said softly. "That explains the size of the satellite. It would take ellipses and ellipses for us to extract all that information and try to put it in any sort of order. We will have to just pick and choose the type of data we want."

"What, exactly, do we want?" asked Lise.

"A better question is how much data do we want," said Eve. "It is very likely that any data we access will just raise more questions that will require us to access even more data. That's a hole we don't have time to fall too far into. We will have to focus on the data that is most crucial to our mission."

"I'm not sure I understand," admitted Lise.

"And that's fine. You are still technically a student," said Eve reassuringly. Eve took a deep breath and appeared to lose herself in the data stream.

"What is it?"

"There is a big part of Umae history that we don't have any information about. The Cataclysm, the Directors, the origin of the Protocols. All of that information might be in that satellite. If it is, and if we can access it, we might be able to determine just who, or what, is waiting for us on Haven. We might also be able to tell if the Protocols as they have been handed down to us by the Chroniclers are accurate. If they aren't, and if we have significantly deviated from them, we may have to make changes to our mission plans."

Lise again struggled with the tension between her confusion and her desire to impress Eve. "Why would the Chroniclers have misrepresented the Protocols?" she said finally. Despite the obviousness of the inquiry, it was a question bordering on blasphemy.

Eve did not seem concerned. "Umae error for one. We Journeyers have no information at all as to how well the Chroniclers utilize the technology for accessing information about past events. We only know that they try and that they claim an ability to do so. But the fact that they don't share much detail about those things has always been a concern. Min….." she paused for a second and took a breath, "….Min wasn't convinced they knew much about anything. He had always wanted to study the logic of the Protocols in depth but never really had the opportunity." She turned towards Lise. "We might have that opportunity now."

Lise nodded. "So where do we start?"

Eve placed her hand on the display and manipulated the data schematic in front of them. "Look at these," she said pointing to the display. There was one bright cluster of light that was larger than any other figure on the display. A second cluster, of identical shape to the first but much smaller, was in the far corner of the display. "These are both data matrices addressing the Cataclysm. The amount of data in each matrix is vastly different. The larger one has so much data in it that I'm not sure our extraction protocols could give us a meaningful analysis within a reasonable amount of time. The second one is very small. I am wondering why there is such a great discrepancy since they address the same thing."

"Can we use the extraction protocols to analyze the smaller one?" asked Lise. "Maybe that would at least give us some ideas about what sort of data is in the larger one."

Eve nodded approvingly. "Exactly what I was going to suggest. Well considered." Lise beamed. "I'm going to initiate the extraction and analysis of the smaller batch now."

Once more, Eve manipulated the data on the display and then engaged the panel to extract and analyze the smaller data batch. The panel completed the process almost instantly.

"Well?" asked Lise. The various shapes representing data batches had been replaced by simple text. It didn't take Eve long to read through it. As she finished she sat back in her chair. An expression of weighty disbelief washed over her as her shoulders slumped.

"I.....can't believe it."

"What? Tell me!" Lise started to review the text herself.

"It's a history of the term. A history of the word 'Cataclysm'," said Eve finally. "The word......it's only been in use for about twelve hundred ellipses."

"So?" asked Lise as she continued looking at the text. "Why does that matter?"

Eve took Lise gently by the arm. "Lise, the Directors gave us a calculation for Critical Closure that began the countdown at least ten thousand ellipses ago. We always assumed that was around the time the Cataclysm took place. We also assumed that the Cataclysm was the reason for the Protocols in the first place, that they were implemented to prevent a second Cataclysm."

By now Lise had abandoned all concerns about disappointing her mentor. "So the Cataclysm happened more recently than we thought. What does that mean?"

"No," said Eve, "you don't understand. We have data in the *Starshine's* logs dating back far, far beyond that. By thousands of ellipses. That was primarily the reason we assumed the Cataclysm happened when it did. But if the very word 'Cataclysm' is only twelve hundred ellipses old, that can't be true."

Lise turned away from the display, trying to make sense of what Eve was saying. "How could there be a word for some historical event that happened over ten thousand ellipses ago that is only twelve hundred ellipses old? Where did that term come from?"

Ordinarily Eve would have been pleased with the connection Lise was making. "The Chroniclers. They created the word. But they did it almost eight thousand ellipses after it supposedly happened."

"'Supposedly'?"

"I think it is very likely that instead of just recording history, the Chroniclers attempted to guess at what happened and then passed it along to us as accurate."

"But why would they do that?" Lise turned back to the text on the display, trying to find another possible explanation.

"The Chroniclers are in a position of authority. But that authority derives from the perception that they understand our history and are the keepers of the path to our salvation."

"Wait, wait. What perception? By who?" protested Lise.

"Ours," said Eve slowly. "By the Journeyers. We believed the Chroniclers understood our history. That they knew all about the Directors and the Protocols, but for some reason unknown to us they just couldn't tell us everything. But......"

"They lied."

Eve nodded. "It would appear so. Min was right to have doubts about them. Every Journeyer before him considered such doubts to be blasphemous." Eve reactivated the extraction protocol and the various shapes representing the various data matrices they had seen before reappeared on the panel. The huge shape representing the other batch of data concerning the Cataclysm dominated the screen once more.

"So, what do you suppose is in that batch of data?" asked Lise.

Eve paused. "Everything. The truth. About the Protocols......."

"And the Directors?"

"Yes. And the Directors."

Lise stared at the figure on the display, half expecting it to start moving about. "That's why it is so big."

"Exactly."

"Too big to examine with the extractor protocols?"

Eve released a long, deep sigh and stood up. She stepped back from her chair. Keeping her eyes on the display. "We only have so much time. I'll have to do it myself."

"What do you mean?" Eve's concerns were beginning to weigh on Lise as well, even though she didn't entirely understand them.

"I'll have to connect to the data using the neuro-tactile interface. Once I prepare my mind I should be able to find the most relevant data in the shortest amount of time."

Lise jumped up from her seat and interposed herself between Eve and the panel. "No! You can't! There's far too much data! Your mind will never be able to process all of it!"

Eve glowered at her student. "You don't give me enough credit. I can do it. It's the only way. It's the only way we will finally know everything."

"But what if you.....can't?" Lise was breathing hard and she felt the tears welling in her eyes.

"I have to believe that I can," said Eve. "Or I won't be able to. So shush." She began gathering some small accessories from atop the panel. "I'll need time to prepare. I'd appreciate it if you would run diagnostics on the neuro-tactile buffers. Every little bit will help."

Lise nodded. "I will. And I will," she replied, trying to sound confident. "And I'll be here ready to help you, any way I can."

Eve grasped Lise gently on the upper arm before turning and walking off towards her chamber.

CHAPTER 7

"Joba! Joba!" An Umae woman called out to either side of the trail, waiting for a response that didn't come. She stopped to lean on her staff, looking upwards to gauge how much time she had before the bright, boiling sphere appeared in the morningside sky. A rustling noise off to one side of the path drew her attention.

"Any luck?" A Hek male exited the brush in front of her. He was thickly muscled and carried a long spear decorated with feathers. His speech was gruff and his pronunciation wasn't perfect, but the woman was encouraged by his progress.

She shook her head. "No, Gondiar. No luck. I know Joba can be headstrong, but he isn't yet of age. I just don't understand why Alitus would have let them stray so far away from the rest of us."

"The prey wander.....far, Pol," said the man, searching for the right words. "Not easy to catch. But we find. Maybe walk with Sky Fire?"

The shadow on Pol's face remained and she again glanced towards the horizon. The full moon provided them with a substantial amount of light, but their visibility was less than perfect. "Perhaps. We might do better if we face it this rotation. We will be better able to find traces of them. This is not like Alitus. Even he isn't that reckless."

"Reck-less?" asked Gondiar.

Pol pursed her lips. "That means.......he would be more careful."

Gondiar grunted. "Yes. He would be......."

A bawling sound rolled out of the overgrowth from their heartside. Gondiar held a protective arm out in front of Pol, interposing himself between her and the noise.

Pol was listening closely. "What was that?" she asked quietly.

"Not know," said Gondiar. "I go........"

The noise swept past them again, this time a bit louder.

"It sounds like something's in pain," said Pol. She gripped her staff a bit more tightly.

"But not Joba," opined Gondiar.

"No, not Joba. But if something is suffering, we should look."

Gondiar considered Pol's suggestion and, rather than seek a perfect interpretation, simply pointed his spear forward and began walking in the direction of the sound. Pol stayed close behind her muscular companion.

The overgrowth beyond the trail consisted of leafy bushes and tall, thin grass. There was no sign that anything of any size had passed through recently enough to leave a trail in the plant life. As Gondiar waded ahead creating a path for Pol, the bawling sound grew louder. Soon they reached a clearing of sorts where a large tree had fallen. They saw a figure huddled up against the tree's trunk. It appeared to be a child. It was naked and was guarding its head with its arms.

"You know?" asked Gondiar.

Pol shook her head. "Hard to say. I can't see his face. And where are his clothes?"

Gondiar sniffed at the air. "Not know. Gondiar look. You stay." He started to creep closer towards the figure next to the tree trunk. Each step he took caused the figure to wail louder as he approached.

"Gondiar, stop!" called Pol. "I think you are scaring it. Come back and see what happens."

The Hek stopped and slowly backpedaled to where Pol waited. As he did, the wailing figure grew calm. "Not hurt," protested Gondiar.

"Oh, I know," said Pol. "Maybe......I don't know. Maybe it isn't familiar with the Hek? Let me try."

Pol slowly eased forward as Gondiar watched with great interest. The figure by the tree watched her approach but remained silent. Encouraged, Pol began to hum a lullaby she used with the younger members of the clan. She made it to within her staff's length without any more protests. She was able to confirm that it was, in fact, an Umae. It was a naked boy of approximately eight ellipses. He continued to cover his face.

"It good?" called Gondiar.

Pol waved back to him and then knelt down, setting her staff next to her. "Are you all right?" she asked in a quiet voice. "My name is Pol. Can I help you?" She leaned in, trying to see his face.

"I know who you are," said the figure in a low voice. "Tell your friend to move away."

Pol tried to concentrate. Her field of vision was completely filled with the sight of this child huddled next to the tree. His voice was her only thought.

"Gondiar, move back," she said in a trembling voice.

"It bad?" asked Gondiar. He began to move forward. The Umae boy let out a brief wail.

"I'm scared," said the child in the same deep voice.

"He's scared," said Pol. "Move back."

Gondiar stopped, once again sniffing at the air. With an unhappy grunt, he retreated until Pol and the figure were just at the edge of his sight.

The figure giggled before removing his hands from his face.

Pol attempted to focus on the child. That face.....it was familiar. It was..... "Dom?" she asked finally. "But how........?"

"Silence," whispered Dom. His command caused Pol to immediately fall quiet. She continued looking at his face, trying to remember. "Pick up your staff." Despite the low volume, his voice was a roll of thunder in her mind. She picked up her staff from the ground and clinched it tightly. "Ask your friend to come over here. When he does, you are going to kill him."

Pol's eyes grew wide momentarily. "But........" It was all she could do to muster that meek protest.

"No," said Dom simply. "You will call him over, and then you will kill him."

Pol closed her eyes tightly but Dom's face persisted in her mind. She opened her eyes again and then stood up and turned around.

"Gondiar, I think it's fine now. Please come." Her voice was cold and flat.

Dom could see Gondiar approaching. "He's quite a bit bigger than you are," Dom noted. "And I'm sure he has much more experience in combat than you do. Do you have a plan? Because you *must* kill him." A dark smile slowly arose at the corners of his mouth.

Pol took a handful of steps forward, her staff in one hand, as Gondiar approached. He was intent on discovering as many details about the figure by the tree as he could. "Is it boy?" asked Gondiar. "Is it good?" He walked past Pol so he could see the child better. "No scared......?"

Pol brought her staff down hard on the top of Gondiar's head. It was unclear if the resultant crunching sound was from the staff or Gondiar's skull. He stumbled ahead slightly as he turned, blood pouring down into his eyes. Instinctively, he raised the tip of his spear and pointed it towards Pol.

"No, no, no!" said Dom in a sing-song voice. He seized the butt of Gondiar's spear and pulled it from the Hek's hands. Gondiar turned in surprise just as the second blow landed squarely on his cheek bone. He staggered and dropped to one knee. Pol began to rain blows down on him as he unsuccessfully tried to cover his head defensively. Again and again crunching echoes rebounded from the nearby trees. As Gondiar swayed on his knees, his hands finally dropped completely. The staff caught him at the temple causing Gondiar to lurch sideways before toppling to the ground. His head was split wide open. Pol looked down at him, her breath coming is gasps. "Well?" asked Dom, as he looked at Gondiar's fallen body. "Did I say you were finished?" Pol's mind spun as she attempted to process what was happening. Regripping her staff, she repeatedly brought it down on Gondiar until her arms could no longer be compelled by even Dom's forceful commands. She slumped to the ground, barely able to breathe.

Gondiar's broken remains were barely recognizable as a Hek. Already the flies were buzzing about, feasting on the blood and gray matter strewn about the scene. Slowly, Pol reached up and wiped away a bit of the carnage that had splattered on her cheek. "N-now.....?" she asked weakly.

Dom let out a low laugh. "Just look at what you have done," he said suggestively. "Look!" Pol shook her head as the fog lifted from her thoughts. She felt a wave rise up from her stomach as she saw Gondiar's body on the ground in front of her. Her hands trembled wildly as she squealed in revulsion. Her staff dropped to the ground.

"W-what did.....I do?" The words choked her and her face grew pale.

"Exactly what I told you." Dom giggled again. "Catch your breath. Because after you do that, we are going to catch up."

Dom had no problems locating the rest of Pol's clan. They had set up camp inside of a large cave on the side of a rocky rise at the edge of the grasslands. He arrived just as the sun began to rise above the morningside horizon. He could see the remnants of their fires and smelled the faint odor of cooked meat. A few of the fires still released weak plumes of smoke into the morning air. But there were only two people outside of the cave. They were both Hek and they huddled underneath a large stone canopy that hung over the opening to the cave. The cave opening faced the morningside sky and as the sun rose the two figures slowly moved farther into the interior of the cave. Dom remained concealed in the tall grass nearby so he could observe this odd behavior.

Once the sun had fully cleared the edge of the horizon its light beamed directly onto the mouth of the cave. The figures had completely retreated inside as the angle of the sun had deprived them of any shade near the exterior of the cave. As the sun passed its zenith the shadow of the canopy slowly crept away from the cave's opening and two figures returned. They constantly shielded their eyes from the light and remained inside the perimeter of the shadow. Dom noted that the people at the opening were replaced by new figures. One Hek and one Umae now stood in the shadows where the original pair had been. Glancing around the immediate vicinity, Dom detected no threat that would warrant such vigilance. The sun's rays poured over his still naked body, prodding his hunger. His manipulation of Pol had done nothing to provide him with any relief from the constant longing to reproduce. His rage at Min returned with full force, along with his hatred of Min's people. Vengeance might not bring him the respite he so desperately craved but he would have it, nonetheless.

Once the sun began to move behind the rise containing the cave's mouth the shadows flourished and the activity around the cave opening increased. A number of figures, a combination of Umae and Hek, peeked outside and muttered quietly among themselves. As the sun faded from view, a stream of people emerged into the quarter light of the evening. Many of them bore spears, slings, and other implements for hunting. Others moved to the fire pits and began working to restore the flames. Umae children scrambled around on the rocks, giggling and laughing as they chased one another.

Dom grimaced. This would not do.

He rose from his place of concealment and began walking towards the cave. The people minding the fire pits noticed him first and called to the others notifying them of his approach. A crowd from within the cave joined those who were outside and moved over to greet Dom. Joba stood in the midst of the group.

"Dom, you're back!" he said enthusiastically.

Dom feigned excitement. "Of course! I told you that I wouldn't abandon you." He saw a few of the children watching him closely. "Can someone offer me something to cover my body?"

Eventually a young Umae woman approached him with a handful of woven cloth. Once Dom unfolded it, he realized it was a covering prepared for a child. By his estimation, it would fit almost perfectly.

"Did you just arrive?" asked Joba.

"Yes, I have traveled a long distance. I'm working to assure your safety and success."

Joba approached him slowly, cautiously reaching out to touch him. Dom made no effort to impede him from doing so.

"Your skin."

"What about it?" Despite having practiced a more tolerant tone of voice, Dom still struggled to conceal his disdain for these creatures.

"It's still white," Joba noted as he ran his fingertip along Dom's arm. "How can that be?"

"What color is it supposed to be?" asked Dom. He was surveying the reactions of the others in an attempt to get ahead of Joba's thought process.

"You have been outside during the entire rotation?"

"Yes. Of course. I told you that I have traveled a long way."

Joba turned towards one of the Hek men who was standing nearby. It appeared to Dom to be one of the men who was guarding the cave entrance earlier. Joba began grunting to him in the Hek language. Dom focused on their exchange.

"Your skin is white," said Joba finally. "Tarak knows that the sky fire was strong during the last rotation. Your skin should be red. Why isn't it?"

Dom realized that all of the Umae around him had pale skin. The Hek had skin that was only slightly darker. The behavior he had noted earlier made sense now. They all avoided the sun.

Dom finished donning his new clothes. "I told you," he said firmly, "I am a Journeyer and am capable of many wonders." He nodded towards Tarak. "Tell him." Dom had grown comfortable with the Hek language but wanted to be sure he spoke it perfectly before presenting it to this group.

Joba turned towards the Hek man and repeated what Dom had just said but in the Hek language. Tarak's eyes widened and the Hek listening nearby rumbled with awe. Tarak offered Joba a new question. Joba returned his attention to Dom.

"Tarak wants to know........"

"More than you can count," said Dom. He offered his response in the rough tongue employed by the Hek. "You do not have a number for the rotations I've been wandering this world. It is like the blades of grass in the field."

Another louder rumble rolled through the crowd.

"You can speak Hek?" asked Tarak. The warrior's eyes narrowed as he assessed this strange child.

"I can," said Dom plainly. "As I told you, I am capable of a great many wonders. And I will show them to you. Maybe," he added with a slight edge to his tone, "more than you'd like to see."

The entire clan was gathered inside the cave complex. There were perhaps eighty individuals of which seventy-five percent were Hek. There were a few of the Hybrids Dom had seen on his travels and they seemed to readily interact with both the Umae and the Hek. A few dogs lounged lazily around the interior of the cave. Light poured into the cave's opening but no one remained too close, preferring instead to remain in the shadows away from the Sky Fire's rays. A few small fires burned throughout the area. Dom sat in the middle with every clan member intently listening to him speak.

"I would know of the problems that you have," he said firmly in perfect Hek, his voice filling the chamber. "In order that I might solve them. It does not appear that many Umae survived the Great Rain. Tell me, are all of you from the same settlement?" The heat of the sun and its arousing effects had hampered his judgment. While he had recognized Pol he had not considered the possibility that others from Atla had survived. Any survivors might possibly remember him from before the Pulse.

Tarak stepped forward. "The Umae came from many places," he explained in Hek. But none from the island survived the Great Rain." The Umae in the room listened carefully to Tarak, but Joba nonetheless interpreted his words so everyone would fully understand what was being said.

"You are a translator?" Dom asked Joba.

"Yes," said the young man. "My mother taught me the Umae tongue and I grew up around the Hek. I am one of the few who can speak both tongues very well. I have tried to teach everyone to speak both, but there is only so much time."

"Your mother," asked Dom, resisting the urge to smile, "who is she?"

"She is called 'Pol'. But she is not here. She is......still looking for me apparently." Dom noted a taint of worry staining Joba's voice.

"When do you expect her?" asked Dom. He was confident that his feigned concern sounded authentic.

"She should have returned by now. But she is with Gondiar and a group of capable Hek trackers. I'm sure she will be here soon."

"I'm sure you are right," noted Dom dryly. He turned back to Tarak. "You are a leader of some sort?"

Tarak glanced at Joba before nodding. "I am a clan elder. But there are others."

"Yes," came a voice in the back. An Umae woman came forward, the others parting to let her through. "I am one of the others. I am called 'Nagham'." Despite her race she was speaking Hek.

"And which settlement did you come from?"

"I am from the island. The clan history says I was a gift to the Hek from the Umae. The Hek are my people."

"A gift. How interesting. So how many clan elders are there?"

"We are as fingers on a hand," answered Tarak. Joba continued to interpret for the Umae who were trying to listen. "Tarak, Pol, Nagham, Liswan and Gondiar. Gondiar is with Pol."

"And where is Liswan?"

"Liswan is hunting. When the Sky Fire appears, we seek shelter. He will not return until after the Sky Fire runs away."

Dom looked at the light entering the cave and considered Tarak's comment. "Why do you avoid the Sky Fire?" he asked. "It is a source of great power offered to those with the courage to use it."

"No," said Nagham. "The Sky Fire is a torment. It blinds us and burns our skin. It has driven the clouds from the sky. Where once water was freely offered by the sky, we now must search for it in the darkness. The Umae have shared tales of the Journeyers. If you are truly so wise, how can you not know these things?"

Dom bristled. "I have walked underneath the Sky Fire without being burned. I already know both of the tongues spoken by your clan. Do you still doubt my wisdom?"

Following Joba's translation, the entire crowd murmured to one another, briefly debating Dom's point. "You claim to be the bearer of gifts," said Nagham. "So far, you have brought us only words. You have asked for us to tell you our problems, yet also claim great wisdom. How is it that you do not already know?"

Nagham was tall for an Umae woman and she covered only a small portion of her body with furs, just like the Hek. Her skin was very pale and her hair the color of milk. Her light red eyes focused confidently as she awaited Dom's response.

Dom already knew what her fate was going to be. "The Hek have been cursed," he said at last. Joba's translation nearly sent the group into a frenzy. "Look around. I see young Umae, but I see no Hek with fewer ellipses than Joba. None have been born."

Nagham drew her shoulders back. "This is true. Since the Great Rain, no Hek have been born. The Sky Fire is our enemy. It has taken our children."

"But not all of your children," noted Dom. He pointed to a couple of women each cradling an Umae baby in her arms. "Only the Hek children. Why do you suppose the Sky Fire has done this?"

Tarak stepped forward and had a brief, quiet consultation with Nagham. "We do not know," Tarak admitted finally. "But you said you have great wisdom and can help us. Perhaps this would be a way for you to prove these things."

Dom nodded with false humility. "Yes, I can help you and I will. But first," he turned towards Joba. "You are worried about your mother and those she is with. Perhaps we can find her. It will be best if we have all of the clan leaders here at once, don't you think?"

After a moment, Nagham nodded and Tarak followed her lead. "Tarak will go search," said Tarak who now spoke Umae. "Joba and Dom come, too. Tarak know which direction they go. Will not be hard to find."

"Agreed," said Dom. "When should we leave?"

"Nagham shall go as well," the Umae woman insisted. "But not until the Sky Fire has run away. Pol and the others will be sheltered until then anyway. They will be easier for us to find if they are awake and moving."

"Very well," said Dom. "As soon as the Sky Fire is gone, we will go." He followed Nagham's movements with great interest. *And my plans to rule this clan will begin with you.*

CHAPTER 8

Eve and Lise had finished the final preparations for Eve's interface with the satellite. Eve was wearing a simple, loose robe and had coated her hands with a lotion designed to optimize the transfer of data from panel to Data Agent. She stood in a doorway just as Lise was completing an interface with the panel.

"The interface's functions are entirely within the safety parameters Eve requested." It was Min's familiar voice, and one that caused a warmth to rise in Eve's cheeks. Even when speaking about safety parameters, the sound of his voice saying her name caused a fluttering in her stomach.

"Thank you, Min," said Lise. "I'm worried about this procedure. I'm sure you've made it as safe as possible, but still......"

"Eve is a highly skilled Data Agent," said sim-Min. "In fact, she taught the real me quite a bit about the topic. Her record of success is unblemished. She has an excellent probability for success."

"I know she does. She is very important to me," said Lise quietly. "I want her to be ok." Eve closed her eyes, hungry to hear sim-Min say what she wanted to hear.

"As do I," said the voice from the panel. "I look forward to reviewing the report from the process after it has been completed. If you are finished with me, I will go."

No! Don't go!

"Fine," replied Lise. "And thanks again."

Eve took a moment to compose herself, his voice still fresh in her mind. "Ready?"

Lise looked up. "Oh. Yes. Min and I, well sim-Min and I, were just reviewing the safety specifications. He says they are in line."

"Then I'm sure they are."

"I thought he was a Tech Agent before he was the Command Agent," said Lise. "He said you taught him about being a Data Agent too. Isn't there anything he can't do?"

Eve stiffened. "He was quite talented. I showed him a few things about interfacing with data batches using the neuro-tactile interface. He helped to dramatically advance the technology, so it only seemed right to show him some of the preparations necessary to use it."

"I'm looking forward to meeting him someday," said Lise. Eve turned away, feigning a coughing spell. "Are you all right?"

Eve's eyes were red and watery and her cheeks were flushed. "Yes," she said, taking a couple of deep breaths. "Fine."

"What's next?"

"I know we've been over this," said Eve. "But it's worth repeating. Besides, we can consider this a part of your education. A rather advanced part."

Lise was fascinated by the colored shapes representing the data matrices that were on the display. The central shape representing what Eve had determined to be the entire batch of data concerning the time period that ended 1200 ellipses ago loomed directly in the center of the panel.

"It's so big!" lamented Lise. "Please, there has to be another way. I've seen the tutorials. This is too dangerous."

"I wrote the tutorials," noted Eve. "And it's only dangerous if you don't fully understand what you are doing."

"But we could use the panel to do everything!" continued Lise. "There wouldn't be any danger at all."

Eve was beginning to grow annoyed. "Lise, we've been over this. Algorithms are fantastic tools. And if we had all the time we wanted, that's how we would do this. But this data batch is very large, and very complicated. It would take more time than we can spare to download all of it. There's no adequate way for a machine to judge what we need and what we don't. I've prepared my mind to make those judgment calls subconsciously. And if it turns out that my mind doesn't immediately recognize a particular fact as important, we can always recall it from my subconscious later." She noted the continued expression of anxiety on Lise's face.

"And you go into stasis afterwards?"

"Exactly. The repose chambers can induce the appropriate level of sleep for optimal processing by my brain."

"How long does that last?"

Eve shrugged. "It's hard to say. I have never had to remain in stasis for more than half a rotation." *But I want to be in stasis now so my brain will be quiet.*

"But the danger! It's not a perfect procedure. You could be injured. You could go insane. You could........"

Eve forced a smile. "I could execute the procedure flawlessly and provide us with access to an invaluable amount of data that we've never had before." *And know if all of this was worth it.*

Lise was not relieved. "What do you want me to do?"

"First and foremost, don't interfere."

"But......."

"No 'buts'!" insisted Eve. "Whatever happens, do not break the connection between the panel and me. That has to be done by the Data Agent. Otherwise, the risk of damage to the Data Agent from the feedback goes up dramatically. I want you to watch, monitor, and learn." She turned her attention back to the panel. "Everything will be fine."

"Min said the same thing."

Min wasn't always right.

Eve rubbed her hands together slowly and closed her eyes. After what seemed like an eternity to Lise, Eve finally stretched her hands out and placed them palms forward onto the face of the display in front of her. After one final deep breath, she leaned in towards the display, her face etched by concentration.

The shape representing the target data began to grow and quickly pushed all of the other shapes off the edge of the display. It then began to slowly change colors. It passed through the shades of blue, and then of green, and finally of red until it settled on a shade of deep crimson. Eve's face tightened and her lips pursed as she concentrated on the energy being passed to her mind by the panel.

Without sound, the shape on the panel exploded into a cascade of vivid colors. Eve's back arched and she moaned quietly, but she maintained contact with the display. Sweat quickly gathered on her forehead and began dripping down her face onto the panel in front of her. It was all Lise could do to resist pulling Eve away from the display.

There didn't appear to be any pattern to the arrangement of colors playing on the field in front of Eve. Bands from every part of the visible spectrum writhed and twisted about one another. These knots then spun rapidly before colliding with one another sending "sparks" of new colors zipping across the display. Eve's hands trembled and her lip began to bleed from where she had bitten down on it. The color weave now formed a spinning tunnel that seemed to be pulling Eve forwards towards the display. Lise stood up and flexed her hands helplessly. The tunnel on the display depicted the view of one plunging headlong down a multi-chromatic pipe. As the pipe straightened, a bright yellow glow formed at what would be the "bottom". The glow was another data package and it grew larger and brighter and it accelerated towards the face of the display.

Eve cried out, her voice saturated with terror. Again and again she wailed as the bright yellow matrix rushed towards her hands. She stood up slightly from her chair and leaned forward just as the yellow mass reached its zenith. The display exploded with a yellow burst so bright that it left spots in front of Lise's eye. She staggered back, trying to regain her bearings as she heard Eve release one final, pained scream.

"Eve?" called Lise. Her vision was still impaired by the chromatic echoes of the burst from the display. She moved slowly in the direction where she had last seen Eve. "Are you all right?"

Lise reached out in front of her, finding the back of Eve's chair. She blinked hard and shook her head, trying to rid her sight of the blinding colors. Finally, she could see. Eve was face down on the panel, unmoving. The display was blank. Lise knelt down next to her mentor. "Eve! Eve! What happened?" Eve remained motionless. Lise looked around, panic-stricken. A pair of mechs stood against the far wall.

"Help me!" she screamed. Her voice was nearly choked with fear. The mechs paused briefly, seemingly to assess the situation. "Take her to her repose chamber!" Lise barked. "Now!" The mechs tapped into the command hierarchy in the Alpha Panel. The ship was only crewed by two Umae, and one of them was unconscious. Their protocols permitted them to accept limited direction under that circumstance, even from one without Command Authority.

Both of the machines moved forward, each positioning itself on one side of Eve. The mechs remained silent as they carefully scooped Eve's limp form out of her chair and began moving her down the hallway. The mechs followed the long hallway leading away from the *Starshine*'s Alpha panel before making a turn towards the repose chambers. One of the chambers was already open, and the mechs easily set Eve inside. The chamber's door slid shut and the lights on the inside emitted a gentle blue glow. The mechs reversed course away from the repose chamber and moved off towards the long hallway.

Eve had never considered this contingency to be a significant possibility so she had never told Lise what she was supposed to do now. "Panel, what do I do now?"

There was a large panel along the far wall that activated following her inquiry. "You lack Command Authority," said a cold, monotonic voice.

"But she is hurt! I don't know what to do!"

"You lack Command Authority."

"I heard you!" screamed Lise. "But....." She slumped to her knees, her body shaking with sobs. She watched Eve's lifeless form in the repose tube, trying to will her back. She had to reach the *Aurora*. "Open communications with the *Aurora*," she said, trying to compose herself.

"You lack......."

"Shut up!" Lise pressed her forehead to the cool, metallic floor searching for focus. Focus? Yes. She knew what she had to do. She got up and went off to seek the only source of assistance she might find on the *Starshine*.

Lise hadn't noticed the silence in the chamber surrounding the *Starshine's* Alpha panel. Typically she was there with Eve either going over her training protocols or interacting with some simulated computer profile. It all depended on why she was there. Right now, she needed *him*. Or at least as close to him as she could get.

"Activate simulation profile 'Min'," she said.

The silence was uninterrupted momentarily until she heard the cold intonation of the panel's "voice".

"Command Authority required."

"What?! I don't understand. I was just utilizing that profile a short time ago. I need it again."

"The prior utilization was terminated. Command Authority is required to re-initiate."

"But I never needed that before!" She was raising her voice with the panel, not appreciating that neither her volume nor tone made any difference to it at all.

"Prior source of authority is no longer capable of assigning authority."

She reached out and touched the surface of the panel. The metal chilled the palm of her sweaty hand. The display was blank. "I used Eve's authority," she said. "I want to do that again."

The panel remained silent long enough for Lise to wonder if it had become inactive. It replied just as she began to repeat her last request. "Eve is incapacitated. Her prior authorization is no longer valid."

"I KNOW she's incapacitated!" shrieked Lise. "Why do you think I need to talk to Min???"

"Command Authority required."

"Ahhhh!" Lise slammed her fists against the surface of the panel. She leaned forward and put her hands over her face. Something had gone terribly wrong with Eve's procedure, something Eve herself had considered almost impossible. If she hadn't, Eve would have left a contingency. Without one, Lise could do very little. And if Eve didn't wake up........

Lise glared at the display. What if she could link to the panel like Eve was teaching her to do? Could she find some workaround? But what if she failed and ended up like Eve, or worse?

"So what am I supposed to do? Because I don't know! Eve might be......" the word caught in her throat...... "dying for all I know. And without some sort of operational authority I can't really do anything. Not for her. Not for me. Not for the Umae. But I don't have that authority."

Again, momentary silence. Lise was once again struck by the impression that the panel was thinking.

"But I do." The sound of his voice brought a fresh wave of tears to her eyes.

"Min?" she asked. She was trembling with relief.

"A Min simulation," said his voice. But she couldn't tell the difference. "Can you give me Command Authority?"

"No, I can't do that. But you won't need it. I have Command Authority."

"I don't understand."

"I don't doubt that. I simply needed time to interact with sim-Sol to determine if my override could be employed."

"Override?"

"Let me assure you that Eve is going to be fine. I'm still analyzing the specifics of her interface with the satellite, but my preliminary conclusion is that she was preoccupied by something while she was attempting to execute it. The result was that her mind couldn't adequately control the information flow and screen it properly. It was, neurologically speaking, more than she could stand. So she lost consciousness."

Lise released a long breath. She could feel her heart pounding in her chest. "She's all right," she said quietly. "She's going to be alright."

"Yes. Sim-Sol offered that opinion to a near certain degree. But her recovery will take some time."

Lise sat back down at the Alpha panel. "So what is happening here? I'm confused."

"About what?"

"You are a simulated personality profile but you have Command Authority?"

"Yes, I can see why that might be confusing. I got it from Min. After all, he was the one who designed the directives that implemented all of the sim profiles on the *Starshine*. He simply provided his own simulation with contingent Command Authority if certain parameters were met as they are now."

"Great. So I need some help."

"That was one of the parameters." Lise half-expected to hear the panel chuckle at its own joke.

"What am I supposed to do?"

"I've reviewed the most recent communications with the *Aurora*. Interesting that Jack decided to augment the communications capacities with a simple radio wave-based platform. I have insufficient data as to just how much faster the *Aurora* is than the *Starshine*. I also lack sufficient data to estimate to any substantial degree of certainty the distance to the Haven candidate he described. However, the prior scans would suggest that it is no farther away than one point two light ellipses. That should give us time to do what we need to do."

"Which is what?"

"According to sim-Sol, Eve will likely require a period of convalescence in the repose chamber of at least an ellipse."

"An ellipse! Why so long?"

"She incurred a significant amount of neurological stress. Fortunately, the repose chamber can be integrated with the Life Lab's panels to maximize recovery. According to sim-Sol, neurological injuries of that type take that long to heal. And we also want to make sure her procedure wasn't for naught."

"What does that mean?"

"We could bring her out earlier and, from a simple standpoint of health, she would be fine. However, her subconscious needs time to sort and order all of the data she channeled. If we don't give that process enough time, she won't be able to tell us anything about what she learned. It will have been forgotten."

Lise sat back in her chair and closed her eyes. "So I'm going to be here all by myself for an ellipse and a half?"

"The time is an approximation, but likely a close estimate. But depending on what you mean by 'alone', no. You won't be."

"Because you will be here?"

"I will be here. Sol will be here. Hab will be here. Any former member of the *Starshine*'s crew that you want to interact with will be here. Even Eve if necessary. I'm speaking of their simulations, of course."

"And what will I be doing to pass the time?"

"That's up to you, but I have a suggestion. I've reviewed your training logs. You have almost completed your training as a Data Agent. And you've done very well if I might add. You could finish that and then use the rest of the time learning about anything you want to learn about from the very best teachers you could ask for."

Lise finally allowed herself to smile. "Yes. I could do that, couldn't I?"

"Indeed. In the meantime, I will send a message to the *Aurora*. I will have to send it several times at varying planar angles from our position, but if its reception capacities are similar to the ones Jack constructed on this ship, the message should get through. The array is quite powerful. It will take quite a long time for the message to get there, but at least they will eventually know what our status is."

Lise rose from her chair. "Are you going to leave if I stop talking to you?"

"Not at all. I'll remain until a live Umae arrives with Command Authority and shuts me off again. Min didn't intend to substitute his Command Authority for someone else's. He simply intended on making sure someone with command experience had such authority on the *Starshine* at all times."

"Eve told me that you were very smart. I guess I didn't appreciate just how smart."

"She told you that?" Lise could easily imagine sim-Min blushing.

"She has said a lot of nice things about you. She misses you. Very much."

Sim-Min paused, possibly searching for the right words. "It's good to be missed, I suppose. It means I was noticed to begin with, doesn't it?"

Lise's thoughts drifted for a moment towards Mac as she imagined his strong, silent self holding her hand. "Thank you, Min. You've given me a lot of hope. Just in time, too."

Lise took sim-Min's advice to heart. She knew that unless she found a way to occupy her mind she might eventually go mad for lack of Umae companionship. The various simulations offered by the panel were better than nothing, but after interacting with them for a while she could discern the difference between a mathematical approximation of a personality and the actual person. The simulations couldn't touch her. They couldn't exchange messages with facial expressions. They couldn't sense her feelings and didn't know what to do to make her feel better. But they did serve as the distractions she so desperately needed.

Sim-Eve was almost as good of a teacher as her actual mentor. Lise was able to finish her Data Agent training ahead of schedule and had time to accept instruction from sim-Sol and sim-Min in their respective areas of expertise. Sim-Sol was a particular joy. Even as a simulation, he conveyed a deep love of life and an appreciation of everything living. She felt it might prove most practical if she learned some basic medical skills so she had him teach her about the treatment of a variety of different injuries. She also learned about the basics of disease diagnosis and treatment. Sim-Sol saw these situations as temporary deviations from the normal state of an Umae. While injuries and illnesses were conditions that could be repaired so the Umae could return to full productivity, for sim-Sol it was more than that. Various health issues caused his patients pain and stress. He subordinated the efficiency of the patient to the comfort of the patient. They were Umae first and foremost and that was enough. Whatever it was that they had to offer society was secondary. Lise hoped that she would have the chance to spend some time with the real Sol before all of this was over.

By the time Lise had completed her training and supplemented it with tutelage in areas she had never dreamed that she'd find interesting, more than a full ellipse had passed. There had still been no reply from the **Aurora**. She didn't know where they were or what they were doing. While that troubled her sleep, she couldn't deny that Mac was the one she worried most about. Even though it had been so long since she had seen him, his image in her mind still made her heart flutter. It troubled her to think that perhaps he had forgotten about her. It was just this sort of situation that the simulations couldn't help her with.

She sat in front of the Alpha panel, staring blankly ahead. She couldn't focus and had no idea what it was demonstrating. A voice pulled her back from the fog.

"Lise, you do not appear to be processing any of the lessons on the Alpha panel." It was sim-Min.

"How can you tell?" she asked gloomily.

"Once you have completed a certain segment, you direct the panel to move ahead to the next segment. The rate with which you do that has decreased markedly. Since your test results are excellent, it is unlikely that this is due to an inability on your part to comprehend the material. The next most likely explanation is a lack of effort."

Lise folded her hands in front of her. "I don't think it is anything you can help me with," she replied. "I need to talk to a real Umae." For a moment she feared that her choice of words might offend sim-Min. The simulations were that good.

"That isn't possible right now," said sim-Min. "But I may have the next best possible option. One moment."

Every time sim-Min told her to wait, she imagined that it had to go into another room somewhere and look something up. After a brief time, it would announce its reappearance and then tell her something it couldn't tell her earlier.

"Can I help you with something?" Now she was talking with sim-Eve.

"Not any more than Min could," said Lise. "Wait, did sim-Min go and tell you to talk to me?"

"Sort of." Sim-Eve's tone was lighter and more comforting than sim-Min's. "He consulted with sim-Sol first and then sim-Sol told me to come talk to you."

"I'm not sure I'll ever get used to these simulations."

"You need to communicate with Eve."

"This isn't anything you can help me with," explained Lise.

"I meant the actual Eve," said the simulation.

Lise stood up, a smile blooming on her face. "She's awake? Why didn't I get some notification?" She turned to head towards the repose chamber.

"No, she's not," said sim-Eve. "And Sol says that she won't be for a while. But you can still communicate with her."

Lise sat back down. "How? How do I do that?"

"You have completed your Data Agent training," began sim-Eve. "The repose chambers allow for certain people to interface with an individual inside the repose chamber."

"Certain people?"

"Yes. A Data Agent who has had sufficient interaction with the individual in the chamber can interface with the individual. It is much like exchanging information with the panel. The fact that you are both Data Agents will make the process much easier."

Lise brightened a bit. The chance to communicate with another Umae sounded so appealing. The fact that she could speak with Eve was even better.

"Just like the panel? All right. I'm going to try."

Lise rose from her chair and walked back towards the repose chambers. Her footsteps seemed to raise a cold, metallic echo in the passageway. She stopped for a moment and could hear the faint sounds the ship made as it progressed through space. A subtle hum, the occasional soft "ping" – all noises she was accustomed to. Maybe the echoes of footsteps had been there before as well but had evaded her perceptions somehow.

Eve looked just as she had every other time Lise had come to visit her. Except this time, there really would be a "visit".

Lise took a deep breath and began the process of preparing her mind. Eve needed practically no time at all to prepare for an interface, but she had been doing them for many, many more ellipses than Lise had. This was her first attempt at this type of interface and it was with her mentor and friend. Lise didn't want to mess it up.

She thought for a moment about what she needed to communicate about.

It was Mac primarily. Eve knew what it was like to be separated from one who was dear to her. She would understand and be able to help. Lise was sure.

Finally, Lise made a brief entry into the panel on the outside of Eve's chamber before placing her palm on the face of its display. She felt as if she were reaching out to hug a dear friend......

Lise's body tensed as adrenaline rushed into her system. The connection was made immediately as if Eve's consciousness had been waiting there for Lise to enter. Lise could feel the anxiety pouring from the display into her own mind.

*Lise! Contact the **Aurora**, immediately! Tell them not to land on Haven!*

Lise swallowed hard as the sweat began to bead on her forehead.

What?!? Why?!? Is something wrong?

Yes! The data from the satellite.......

Eve? Eve?!? Are you there? Min can you help me?

..................Min?

The Min simulation. He has been helping me.

Have him help you contact the Aurora! Go quickly!

But what is it? What's wrong? Lise waited but got no reply. *Eve? Eve? Please come back!*

Lise could no longer feel the connection. She tried to re-establish it but Eve's presence was no longer open to her. Lise stepped away from the repose chamber and tried to calm her shaking body. Eve looked exactly the same as she always had.

"Min, what happened?" said Lise finally.

After a brief period of silence, it was sim-Sol who replied. "You shouldn't attempt that connection again until I approve it," it said.

"Is she all right?"

"It would be best if I explained it later."

"But....."

"Lise, please." Sim-Sol seemed to be begging. "Trust me."

Lise looked over Eve's form one more time. "I do," she said finally. "Just don't make me wait too long."

CHAPTER 9

Mac sat in front of his training panel, frustrated by both the information on his display and the tone taken by the sim-Jack serving as his tutor. The young man sat back and held his hands to the sides of his head.

"Mac, come now!" screeched sim-Jack. "You can't be this incompetent forever. Eventually you will be in charge of a Journeycraft. Such a sad reality for your crew."

"Shut....up!" screamed Mac. "I'm doing my best."

"That's what concerns me the most."

Mac stood up and turned away from the panel. "I told you to shut up," he grumbled as he started to walk off.

"Wait. Where are you going? You haven't finished failing yet."

Mac left the main bridge area and headed towards the Life Lab. He stopped at the open doorway and could see Sol and Lin working at a table. Sol was sitting back on a mech that had configured itself into a chair. It could easily move the Life Agent around the lab to the various panels and project tables within. After waiting to be noticed and invited in, Mac realized their degree of focus wouldn't allow that so he let loose with a weak, faint cough. It was enough to draw Sol's attention.

"Hello Mac," he said warmly. "What do you need?" His questions were punctuated by a series of sharp coughs that made Mac feel a bit guilty about his own contrived effort.

Mac entered slowly, unable to meet Sol's gaze. "Um, do you happen to know anything about the thermodynamics of atmospheric adjustment?"

Sol and Lin glanced at one another. "No," said Sol finally. "I can't say that I do. That sounds like Tech Agent territory." He briefly coughed again. "Have you asked Jack?"

Mac drew a deep breath. "He's in his room. All the time. He doesn't really talk to me at all anymore."

"What about the sim profiles?" asked Lin, attempting to sound cheerful. "I'll bet they could be very helpful."

"Yes," Sol agreed. "Try Min's. He was a very patient teacher with Jack. That would prove very instructive."

"Would that one even be available on the *Aurora*?" asked Lin. "Min was never on this Journeycraft."

"It was," said Mac gloomily. "All the sims get transferred to all the other Journeycraft through the tether beam and transfer beam interfaces."

"You said 'it was'?" asked Sol.

"Jack removed it at some point," explained Mac. "He's removed Hab's and every other TA profile that was present, too. Except for his. I'm stuck with it."

Sol's interest was piqued. "What did he say about removing Min's profile, do you remember?"

Mac thought for a moment. "Something about how he wasn't going to have his student receiving instruction from inferior sources. He was very angry."

"'Inferior'," echoed Sol. Lin watched him, trying to catch up.

"Is that important?" she asked.

Sol smiled. "No, I'm sure it isn't," he said. "Just an interesting choice of words."

"Can you just use the tutorials without the sim profile?" Lin was still watching Sol closely.

"I can but the tutorials don't really let you ask a lot of questions. You either get the material or you don't." Mac stared down at the floor. "It's just tough when you aren't getting something."

Lin walked over to her fellow student, placing a comforting hand on his arm. "I know that can be tough," she said. "I struggled with some of the Life Agent tutorials too. But you will get it. It will just click and then you will have it. Just wait."

Mac shook his head. "You had a better teacher." Lin glanced over at Sol noting that the Life Agent appeared pained by Mac's comment.

"Why don't you take a break?" she suggested. "Go somewhere where you can't hear the Jack sim? It's not like Jack will be out of his room soon so there won't be anyone to get after you."

"What do you suppose he is doing in there?" asked Sol.

Mac shook his head. "I have no idea. But when he comes out he's usually very excited. He doesn't talk about it and I'm not about to ask him."

"Well, from a Life Agent's standpoint, I think Lin's suggestion is a fine one. And if Jack does emerge and disapproves, you can tell him it was my directive to you. He may be the C.A., but my decisions concerning the health of the crew still take priority."

"I'll do that," said Mac.

"And Mac?" added Sol in a gentle voice. "I know how much you are missing Lise. Try to stay focused on your studies. That will help pass the time."

Mac nodded solemnly. "Thanks. I hope you are right." He slowly turned and exited through the doorway.

Sol waited a few moments. "Panel, close the Life Lab door." That panel complied, and the door closed with a faint sliding hiss.

"You are worried about something," said Lin.

Sol smiled again. "And you are very, very perceptive."

"What do you think is wrong with Mac?"

"It's not Mac that I'm worried about. It's Jack."

Lin cocked her head. "Oh? Why?"

"Think for a moment. Physical isolation. Mistreatment of other crew members. Superiority complex. Unilateral decision making."

Lin drew a deep breath. "You think he has DSM, don't you?"

"I'll guarantee it," said Sol. "But our options are very limited. He's the only TA we have, not only on this ship but on the *Starshine* as well. He also has Command Authority. There is no way we can complete this mission without him. But….."

"He's dangerous," interjected Lin. "We might not be able to complete it *with* him."

Sol nodded. "Also true."

"So what do we do?"

"The crew is long overdue for their mandated physicals. As Life Agent, I can compel them. We have one convenient element of DSM on our side."

"Which is what?"

Sol hesitated before covering his mouth to cough. "Those with DSM don't think they have DSM. So I don't expect he will object to that part of the exam. It will at least give us enough data to determine the severity of the case."

"What about yours?" asked Lin.

"What about mine what?" laughed Sol.

"Your exam?"

"That's simple. I'll let you do it. In fact, why wait?" He patted his mech chair on its side. "Exam table, please."

The mech rolled across the floor to the nearest exam table and elevated Sol's body until he could easily climb atop its surface. Lin opened a drawer along the edge of the table and removed a metal circlet.

"You are sure you want me to do this?" she asked hesitantly. "I've never done this on a live Umae."

"You'll do fine," assured Sol. He laid back on the table and closed his eyes.

Lin studied the cylinder briefly before sliding it over her hand and up her forearm. An opaque membrane covered her hand as the cylinder locked itself in place just below her elbow. Focusing on the techniques she had learned during her studies, she closed her eyes and took Sol's hand in hers.

At first she could feel the warmth of the old man flow up her arm. Even his physical energy was comforting. She directed her concentration to various aspects of his anatomy and physiology. His nervous system hummed with life.

Her brow furrowed. "Wait…" she gasped slightly. "No." She frowned and opened her eyes before releasing Sol's hand. "I'm sorry, I've made a mistake somewhere."

Sol sat up slowly and took her hand again. "No," he said quietly, "try again."

Lin stared into his eyes, her own wide and filling with fear. Pushing that aside, she refocused and concentrated on the messages the membrane was delivering from Sol's body. His electric field. It was off. His respiration was impaired by……something. Then her eyes grew wide and she backed away from the table.

"No! No!" she protested. "That's a… a… mistake! I did something wrong!"

Sol shook his head. "You are very talented," he said in a hushed voice. "You are correct."

She stepped forward and threw her arms around his slender neck. "How long?" she sobbed.

Sol coughed weakly. "I've known since before our last repose sleep. I considered not taking that risk, but decided I had no choice."

Lin released him, keeping her hands on his shoulders. "I'm not going to let this happen," she said defiantly.

"Easier said than done," chuckled Sol. "I've been trying to cure RTS for the better part of my life."

"But we are going to build on that. After all, now we have another study subject. You."

"Yes, I suppose we do. But I intend to employ that research as a part of your studies. Maybe we can kill two birds at once."

"We aren't going to kill anything." Lin fought back her tears and embraced Sol once again. This time he returned her hug. "After all, we are Life Agents. Right?"

After several rotations, Jack finally emerged from his secret chamber. Although his hair was mussed and he looked as if he hadn't cleaned himself for quite a while, he was beaming. He bounded out into the area where the main panels were finding it empty.

"Mac! Mac! Where are you? I have work for you!" After a brief wait that didn't result in a response from his student, Jack jogged down the passage towards the Life Lab. He found that door to be closed so he rapped loudly on it rather than overriding its locking directive. It opened promptly, allowing him to see Sol and Lise standing next to the primary Life Lab panel reviewing data.

Sol looked up from his seated position on the mech. "Well. I was wondering if I was going to have to scan the ship to see if you had died," he noted.

"Hardly, old man," snapped Jack. "Where is my student?"

"If you are talking about Mac, he's been taking leave the last couple of rotations."

"Leave? I didn't tell him he could do that!"

Sol guided the mech away from the panel in Jack's direction. "You didn't need to," said the Life Agent. "It was on my recommendation. His stress levels were elevated and I thought he needed a break."

"You......you did?" asked Jack incredulously. "YOU decided to give MY student a break?"

Sol nodded. "We just finished with the crew physicals," he explained. "They were overdue. Yours is the only one left."

"I'm fine," said Jack dismissively. "There is no time for that. Now, where is Mac?"

Sol sighed. "I told you. He's taking leave. He needs a couple of more rotations. Then he should be fine."

"I don't care if he's fine," Jack said. His pale white face was growing red hot. "Now. Where is he?"

Sol pointed at a nearby examination table. "Sit down. We need to do your exam. Now."

Jack spun, turning towards the door. "Hardly. Now, I'm going......"

"You aren't going to do anything, Jack." The Command Agent stopped and turned back towards the lab. "Except sit down."

The younger man let out a slow exhale. "You forget, I am your Command Agent. And that is by merit. You do not tell me what I must do."

"I do when it comes to crew physicals," noted Sol. "Lise, please prepare the examination table." Lise hesitated, frozen in place by Jack's glare.

"I don't think she's listening," said Jack with a smirk. "Smart girl."

With a pained expression, Lise walked across the lab and began gathering the necessary implements.

"Jack.....," began Sol in a gentler tone.

"Enough!" said Jack. "I don't have time for this!"

"If your Command Authority were to become suspended, you'd have plenty of time," said Sol.

"You wouldn't dare," snapped Jack.

"If you don't get over to that table for your physical right now, I will consider you unfit for duty. Even your Command Authority is secondary to my purview as ship's Life Agent when it comes to fitness. I won't ask again."

Jack was steaming. He locked eyes with Sol for a seemingly interminable amount of time before finally walking over to the table and sitting down. Sol had gambled on the possibility that Jack had not had the foresight to close off this loophole when he rewrote the *Aurora*'s core directives. "Make it quick," Jack groused.

Sol nodded at Lise and she rolled a small cart over next to the table. Sol took one of the metallic circles from the cart and slid it up to his elbow leaving a thin membrane covering his hand and forearm. "Now hold still."

He stood up from his chair and leaned in next to Jack. Sol cupped Jack's head in his palm and closed his eyes. Jack stared straight ahead as Sol's face became rigid with focus. A thin layer of sweat coated his forehead as he concentrated. He then moved his palm to Jack's face, then his chest, before interfacing with nearly every part of his patient's body. He abruptly removed his hand and slid the metallic ring from his arm.

"Well?" asked Jack.

"You are fine," said Sol. "You might want to change your base vitamin supplement, but I can enter that data into the main panel. Other than that, you are in good shape."

Jack slid off the table. "Told you. Now, I'm guessing Mac is in his quarters. Am I right?"

"Like I told you, Jack, Mac is on leave until the next full rotation. You can't override that directive either."

"You seem to think you understand the limits of your authority pretty well, Sol," noted Jack.

"I've been doing this for a long time."

Jack chuckled. "Indeed. Perhaps too long." He turned towards Lise. "When is your training over?"

Lise shrugged.

"She's getting close," said Sol warmly. "She's an excellent student."

"I'm sure she is. But since I can't speak with my student, I'll share my news with the two of you. I have been steadily narrowing my scans ever since we left the *Starshine*. I've found it. I've found Haven."

Sol and Lise were both dumbfounded. "You're sure?" asked Sol finally.

Jack nodded. "Absolutely certain," he said brashly. "No doubt. And do you know what that means?"

"It means a lot of things," said Sol.

"Its location is right in the path of the beacon signal coming from the satellite we found, just as I knew it would be. I've set a direct course. We won't have to follow the signal after our course intersects it. When we intersect it, we will be at Haven."

"How long?" asked Lise. She was hesitant to speak given the exchange Jack and Sol had had, but Sol seemed to approve of her question.

"One point one nine three ellipses," said Jack. "Exactly. That's not an estimate. And I've already directed all of the ship's scanners to focus on Haven. By the time we get there, I will have collected a lot of data about the planet." He walked to the doorway, stopping briefly before turning around.

"Anything in particular that you would like for us to do before then?" asked Sol.

"I don't really care what you do," said Jack matter-of-factly. "I really don't. Just stay out of my way." With that, he exited and the door slid shut behind him.

Lise stared at the door for a moment. "Hard to believe that he doesn't have DSM," she said. "He has a classical presentation."

"He does have it," said Sol as he sat back down in his chair. A short burst of coughing made him lean over slightly.

Lise put her hand on his shoulder. "Can I get you anything?"

"Just more data," said Sol. "But Lise? His case is very severe. He is dangerous to everyone now and he will only become more dangerous. I know I keep you busy, but in your spare time you need to learn absolutely everything you can about that condition. You may well be the one who ends up treating him for it."

"No!" pleaded Lise. "Don't think like that. We are going to crack RTS once and for all. I just know it." She turned and headed back towards the table where they had been working previously. "Besides, I thought there wasn't any treatment for DSM."

"Segregation. Sedation. Termination," said Sol calmly.

Lise's eyes widened. "Termination?"

Sol forced a smile. "Only a theory," he said, trying to sound comforting. "Just remember the credo of the Life Agent."

"Oh I do," said Lise. "The well-being of the Umae always comes first." Sol nodded. "Always."

There hadn't been any sign of Jack for several rotations. He had retreated to his chamber after his exam leaving Mac alone with the challenge of the Technical Agent training modules. The time off Sol had given Mac seemed to do the young man a great deal of good. Although the Jack-sim continued to belittle him at every turn, Mac was better able to block it out and dig through the processes he was supposed to be mastering.

Sol and Lin spent almost all of their time in the Life Lab. Sol had merged their research into RTS with Lin's training as a Life Agent. The disease offered so many complexities that it served as a sound basis for almost all of Lin's life science training. In addition to the data Sol had when he first became a Life Agent on the **Starshine** so many ellipses past, he had added information from Hab and from Min as well. He had always impressed upon Lin the need for gathering as much data as possible for any type of research. The irony that his own condition would contribute to that store was not lost upon him.

Lin was sitting at a panel, working with sim-Eve to analyze all of their data in as many different ways as they could devise. While RTS offered certain commonalities in every case, there were always a number of rogue factors that appeared to be contributing to its development that weren't always present in every specimen. Statistically speaking there was no sole cause. Worse yet, there didn't even appear to be a core group of factors that were obviously most prominent in predicting the development of the disease. Obviously, space travel was required, but not every Journeyer who traveled in space developed RTS. Of those who did develop it some underwent as many as twenty repose transitions while others only required one. It was oblivious to gender and age. There didn't seem to be a genetic predisposition. Lin had churned the numbers in every way that she and sim-Eve could determine. Without a statistical cause, it was going to be a very difficult condition to prevent. Or, in Sol's case, to cure.

She turned and noticed that Sol was dozing off in his mech chair.

"Maybe you should try to sleep," she suggested.

Sol opened one eye and peeked at her. "What do you think I'm doing?" he asked with a smirk.

Lin smiled. She adored this old Umae. She wanted so much to be worthy of his tutelage.

"I meant in your quarters," said Lin. "Illumination settings, temperature controls, you know. Comfort."

Sol opened his other eye and sat up. "Not ready to stop for the rotation yet," he said. "How are you and Eve getting along?"

"We get along fine," said Lin. "But we aren't having much luck with this data. It just doesn't seem to go anywhere."

"Then you are doing as well as I am," said Sol.

Lin took a deep breath before moving over next to her teacher. "And just how well are you doing?" she asked somberly. "I've noticed that you have been coughing a lot more lately. Do we need to check your metabolic pace levels?"

"I already did. They are decaying. But that's what RTS does. No big surprise."

"How did I miss it?" asked Lin.

"Miss what?"

"You said that you have had RTS since at least prior to our last repose cycle. We've been studying the disease together for much of the time we have been on the *Aurora*. Yet, I didn't recognize that you had it until you told me."

"That's because I cheated," said Sol. "I thought it best that no one knew that I had it. I've ingested enough cough suppressants to choke a herd of tusk cattle. One of the few benefits of collecting as many ellipses as I have is that you learn how to be sneaky. But even those suppressants don't work forever."

"No one else knows?"

"No one. And Jack can't even access that information from the Life Lab data stores. If a Life Agent wants to keep certain health information confidential, it is his prerogative. Even if the Life Agent is the patient. At most, he would be able to determine that confidential information exists, but he wouldn't be able to find out what it is."

"What if Jack changed that? It is his ship after all."

"I don't think he did," offered Sol. "He didn't limit my ability to limit his authority as Command Agent. Did you see his expression when I challenged him on that?"

Lin was torn between admiration for the way Sol handled the nuances of his position and the realization that he was dying and she couldn't help him. The next question was one she had to ask but clung fiercely to her lips.

"Can you.......do it again? Repose?" she asked quietly.

Sol reached out and took her hand. She was amazed at the warmth of his skin. "It's all statistics, right? You know that. Besides, we are going to make it to Haven without a need for repose sleep. That was all I was really hoping for."

"But......." Lin looked away but clung to his hand. She couldn't determine if she wanted more for him or just for herself. "But if Haven is what Jack seems to think it is, what every Umae thought it was since…….forever ago, we will need to go back. We will have to go back and bring the rest of the Umae here."

Sol nodded. "And someone will. It will definitely end up being you, or at least a group you are in. It may well not end up being me. There may be some sense in having a Life Agent on either end of the trip. I'm sure Jack has a plan for preparing Haven so it is ready for the others, but I don't know what it is or who it includes. But based on Jack's current state, I am confident that he will want to return to Uma himself. The prospect of all that adulation would be too much for him to resist. Imagine, he might end up being known as the Umae who finally found Haven and saved our entire species. And he's capable of doing that if his DSM doesn't stop it from happening. And since a Life Agent will likely be going with him, it only makes sense for it to be you and not me. That's why I told you to learn as much about DSM as you can. Jack has the ability to be a hero, but he also has the capacity to lose sight of the prioritization of the Umae if he finds something else to better serve his ego. And the Umae must always be the priority. Always."

Lin felt as if Sol had carefully deposited a load of emotional burdens upon her, little by little, so she could handle all of it once he was finished. "I'm not going to let you die," she said finally. "Not from RTS anyway. You are going to see Haven as our people have always dreamed of it. At least if Jack is right about it."

"His intellect is unaffected by his disease. I'm confident he is right."

"Then so am I." Lin finally brought herself to release his hand. "You know, I've heard so many times that life as a Journeyer is about sacrifice. Min and Eve are separated. Mac and Lise are apart. I never thought that would be something I'd have to worry about. I just planned on doing my job and not getting particularly close to anyone. After all. How could I miss what I'd never had?"

"Because you are an Umae," said Sol simply. "It's a basic need. And even though we are stuck out here in space, we won't always be. Live your life. If you just curl up inside your armored cocoon, you may well evade predators. But sometimes you'll have to go out and seek the sunshine too or you'll die."

Lin giggled. "That's very biological."

Sol grinned back. "It's my specialty. But Lin, I've been close to many of my companions over the ellipses. Min and I have been friends for hundreds of ellipses. I was friends with Hab for almost as long. But their figures are gone from my life now. Do I wish I had never happened with them? Absolutely not.

So don't keep yourself from making those connections. Without them, we may as well just send the Journeycraft into space by themselves."

"Well, I didn't do a very good job of that. The cocoon, I mean."

"Oh?"

She put her arms around Sol's shoulders and pulled him in. "I'll never wish that I hadn't been close to you."

More rotations passed by without further word from Jack. Finally Sol felt compelled to run a Life Census on the *Aurora* just to make sure he was still alive behind his mysterious door. The ship's sensors assured the Life Agent that he was. None of the crew could determine a way to communicate technologically with whatever space he was occupying and sim-Jack was predictably of no assistance at all with their inquiries. All they could do was continue their research and lessons offered to them by Sol and the ship's tutorials.

Lin and sim-Eve began to focus on the differential variables offered by the data gleaned from Sol and Min. They had a fairly large data sample from past subjects who had acquired RTS on other Journeycraft. All of that information, just like all of the other data they received from other ships, was shared through the tether beams and the beacon satellites. Lin was researching the possibility that the age of the last three subjects known to her – Hab, Min and Sol – may have something to do with the condition. Of course, as sim-Eve pointed out, the most obvious connection between age and onset was that the more ellipses an individual traveling in space had, the more time the individual had to develop the condition. That hardly seemed to Lin to be a promising direction of inquiry but she was running short of ideas.

Mac seemed to be flourishing. He was finally making connections through the tutorials without having to access sim-Jack for assistance. This created a positive feedback loop for his progress. The more he accomplished on his own, the less he needed to interact with sim-Jack. The less he interacted with sim-Jack, the better he felt about himself. And the better he felt about himself, the more he accomplished. He even began to attempt to find work arounds to the barriers Jack had erected to prevent any information from escaping his personal chamber. It wasn't something he thought had a high likelihood of success but it provided him a sort of succor in the manner it allowed him to push back against his mentor.

Just when they had all fallen into their regular routines and had practically stopped wondering about Jack at all, he appeared from behind his door. As was always the case when he emerged, he practically crackled with energy.

"Incredible!" It was a strange word with which to greet his crew, particularly when he hadn't seen them for so long. But Mac knew that more explanations would be coming forthwith. Jack never asked for any assistance in understanding anything. His appearances were always about showing off some new brilliance he had accomplished. Mac signaled the Life Lab and Lin and Sol soon appeared, the latter riding atop his mech chair.

"What have you been doing?" scolded Sol.

"It is truly amazing," said Jack, ignoring his question.

"What is?" asked Lin.

Jack smiled broadly, no small hint of mania in his eyes. "The data I've been collecting from Haven," he proclaimed. "My ship's sensors and processing capacities have already made several critical determinations." He paused for effect. "I thought you might want to hear about it." Lin and Sol exchanged glances but both chose to remain silent knowing Jack would continue regardless of their replies.

"None of that data is on the Alpha Panel," said Mac.

"Of course it isn't, you imbecile. It's MY data. I'm only sharing it because there is a very small chance that I might need help from the rest of you. Now listen closely."

Mac didn't shrink back as he had historically. Instead he drew his shoulders back slightly and puffed out his chest.

"Haven will require very little modification to make it habitable for the Umae," Jack proclaimed. "In fact, it is already habitable. But I want it to be more than that. It will be environmental perfection."

"What needs to be modified?" asked Sol.

"There is a broad variety of life forms already on the planet. I've started an analysis of those life forms to determine which of them pose any sort of risk to the Umae. There are a number of large carnivorous predators, both land and sea-based, but nothing that is significantly different from the lifeforms on Uma. None of them are particularly intelligent and our technology will protect us."

"What about microscopic life, Jack?" queried Sol. "That's where the biggest risks will be. Pathogens that we can't even see that could prove fatal to the Umae. We won't have any natural defenses against them."

Jack frowned. "I thought of that already."

"Oh, you did? Then why didn't you include your Life Agent in that analytical process? Don't you think that would have been prudent?" Lin had never seen Sol even remotely angry. She could see his cheeks reddening.

Jack folded his arms and leaned back against the panel behind him. "I don't *need* a Life Agent," he sneered. "I fail to see why it is even recognized as a specialty. Life is just another process like combustion or nuclear decay. I can do everything that needs to be done better than you can, Sol. So why don't you take it easy and just devote yourself to your student. Not that I'll ever need her either."

The color in Sol's cheeks faded away. His typical cheer returned to his countenance. "That sounds delightful," he said. "But you need to understand something. I have quite a large number of ellipses and I'm pretty concerned about what Haven has to offer. It would make me feel a little bit less anxious if I could take a look at the data you and your ship have compiled. No doubt it will set an old man's mind at ease."

Jack nodded. "Of course. The data analysis is very complex but I'll be happy to parse out the essential conclusions for you. Anything to help someone else to feel more comfortable, especially one in your condition."

Lin loosed a surprised gasp, but Sol's expression didn't change. "Old age is a challenge," he admitted. "I appreciate your consideration."

Jack turned and started towards the door of his chamber. "I'll transfer the essentials to the panel in the Life Lab. I have implemented a novel interface involving all of the panels on my ship," he noted casually. He stopped right at the door and turned back around. "It's novel because I developed it myself. The sharing of information, should I choose it to be so, is *very* fluid. It seems so inefficient for one panel to contain data that another doesn't have, don't you think?"

"Indeed," said Sol nodding. "An excellent innovation. Well done."

Jack gave a brief, half bow before disappearing once again behind the door. It slid shut with a resounding thud that practically demanded solitude for the man on the other side.

Sol smiled at Lin, noting the wide-eyed expression of shock that was still on her face.

"I need to talk to you," she said finally. "In the Life Lab."

"I'm going to continue to try and see if I can find out anything about what is going on in that chamber," said Mac. "I'll let you know if I learn anything."

"Thank you, Mac. And we shall keep you advised about the contents of the analysis Jack has promised me," said Sol. With that, the Life Agent and his student proceeded back to the Life Lab.

The door to the Life Lab did not make nearly the authoritative sound when it closed that Jack's door had made. Sol had never noticed that before, but now it made him somewhat uncomfortable. Both he and Lin swept the Life Lab with their eyes, remaining silent for some time. An unintentional smile arose on Lin's face when she saw Frell floating innocently in his repose tube. Each time she saw him she had expected him to look different than the time before. But he remained the same, consistently able to make her feel better about the situation the Journeyers were in.

"Does he know?" asked Lin finally.

Sol shrugged. "He may, but I don't really care. What I care about now is that life data from Haven."

"Can you believe that he didn't involve you in that project?"

"Yes, I can absolutely believe it," said Sol. "And for a moment I nearly let my temper get the best of me. But let's not talk about that now. I'm going to wait for the data to be transferred so I can review it. We can look it over together."

"What do you want me to do?"

"Just what you have been doing. Except be sure that you don't fall behind on your DSM studies. They could turn out to be very important."

CHAPTER 10

The Sky Fire eventually finished its flight to the eveningside sky, spreading shadows around the exterior of the cave complex. With the bright light's absence, the members of the clan slowly began to exit the cave and prepare for the night's activities. Hunters and gatherers readied their tools, hopeful of securing more food for their clanmates. Others headed away to nearby water sources. Dom, Joba, Tarak, and Nagham set forth to search for Pol and her group. The clan members carried a variety of weapons. Dom, as usual, carried nothing.

"You should arm yourself," said Nagham. "There are wild animals about at night. It isn't safe."

"I assure you," said Dom coolly, "that my power and wisdom will protect me. I've no use for a weapon."

Nagham glanced at Tarak. "Very well, let us proceed. He is small. Perhaps a weapon wouldn't make a difference."

Tarak led the group in a heartside direction from the cave. They crossed through a forested area broken by a series of small creeks using only torches borne by Tarak and Joba for light. The sky was clear and filled with tiny twinkling dots. The waning moon still provided the keen-eyed trackers with plenty of light. Occasionally they would hear the growls and grunts of unseen creatures but these night hunters decided, at least for now, to avoid this small group of Umae and Hek.

The forest gave way to a grassy plateau. The undergrowth reached up to the waists of the adults with Dom wading through chest-high grass. He seemed undeterred and unconcerned by the landscape.

"Pol go look by lake," explained Tarak. "Not far," he said as he gestured with his spear. "You hunt there, Joba?"

Joba nodded. "At first, but we weren't finding any game. Alitus wanted to walk in the direction of the low river. That was where we found Dom."

"Pol and Gondiar wouldn't have known that," noted Nagham. "They would have expected you to go where you said you were going to go." The edge in her voice was obvious. "So, we will walk towards the lake."

Tarak took the lead, doing his best to break a trail through the grass for the others. Now that they had left the forest, the chances of being ambushed by a daggercat were much lower. However, they had to keep a lookout for the roverdogs which hunted in packs along these grassy regions. They were generally deterred by fire, but not always. Fortunately, they weren't the most subtle hunters as they chirped and whined at one another while organizing their attack. It would give the group a chance to assume a defensive formation with their spears.

The grass grew sparser and the land slowly began to rise. At the top of a hill the group could see the lake stretch out before them, its surface reflecting the various lights from the night sky. Tarak knelt down and surveyed the ground.

"No sign of anything here," he said finally. "But Gondiar follow Joba. Joba, you come this way? They go where you go."

"No," said Joba. "We started in this direction thinking that if we couldn't find any game we might be able to catch some fish. But we saw a small herd of deer and followed them that way," he said, pointing to the group's emptyside.

"Joba, you need to plan better," said Nagham. "I know this wasn't your hunt, but Alitus should have known better. You will never be a successful hunter by just wandering around until you find prey."

"She speaks the truth," said Dom. "The more calculated the plan, the more satisfying the hunt."

Nagham frowned. "You are a hunter?"

Dom contained the satisfaction welling up inside of him. "Of course. We Journeyers have taught the Umae everything. We must know how to hunt as well." Despite his best efforts, a smug grin bloomed on his face.

"Hmmmmm....----" muttered Nagham as she briefly locked eyes with Tarak. "Let us continue."

They followed Tarak to their emptyside with the Hek man stopping periodically to check for tracks. After the fourth or fifth stop, he suddenly grew animated.

"Here!" he said, pointing. "A number of tracks." He studied the pattern more closely. "Tarak sees four – one Umae and three Hek. That must be group with Pol."

Nagham knelt down next to Tarak. "Yes, I believe you are right," she said. "You have keen eyes, Tarak. They can't be far."

The group proceeded more quickly now, eager to find the others. They moved around the base of a high hill when the wind shifted and a rank odor filled their noses. "Ack!" spat Tarak. "Bad meat!" he complained. "But where?" Despite the unpleasantness of the smell, he sniffed at the air to try and determine where it was originating. "Go here," he said finally. He began to jog ahead, glancing back to make sure the others were following.

He led them around to the far side of the hill. A bald area of dirt and rock was broken up by the presence of a few dead trees. They heard a sudden flapping of wings as dark shapes lifted up into the night. As the wind died the smell hung heavy on the still air. They could hear the faint sound of buzzing.

"Deathbirds leave," said Tarak pointing. "Bad meat there." Despite the rising rancid stench, he jogged towards the trees. His torchlight illuminated the scene by the time the others joined him.

Four bodies hung from a pair of tree limbs. Two Hek men were upside down and their torsos were slit open from throat to waist. Dozens of buzzing insects swarmed about the corpses. Strips of flesh had been torn away by the deathbirds. The bodies' entrails hung from the gaping slashes and dangled across the faces of the deceased.

On the next tree hung another Hek and a female Umae. Their arms and legs had been severed and the remains of their limbs were strewn roughly around the base of the tree. The skull of the Hek was crushed almost to the point of it being unrecognizable.

"Polllllllllll!"

Joba wailed and fell to his knees. As his sobs caused him to double over, his face nearly planted into a decaying chunk of thigh near the bottom of the tree. The ghastly assault on his sense of smell briefly overcame his grief and he began retching on the ground. Tarak took him by the shoulders and pulled him away.

"This was no animal!" insisted Dom. "Look at those wounds. They were caused intentionally, and by someone intelligent." He watched Nagham to note her reaction.

Nagham covered her mouth with her hands. "We must honor their bodies before we do anything," she said. "I will gather wood for a fire." She placed a hand on Joba's shoulder. "We will learn what happened. This is a strange sign indeed."

Tarak gently directed Joba away from the scene. Dom remained for a moment, taking in the sight of the corpses. He then went with Nagham to find wood.

Once Nagham and Dom had left the area, Tarak again approached the corpse-ridden trees. He knelt down and crawled about, looking at the ground very closely. Joba waited, seated on the ground some distance away with his face in his hands. Tarak went back over the same area several times as if to confirm what he was seeing. He then slowly moved a short way away from the area and repeated his assessment of the ground. He looked up at the corpses and then back in the direction Dom and Nagham had gone. He then went and sat next to Joba, offering him comfort. Once they had enough wood, they would burn the bodies. It was the way of the clan. But Nagham, the Gift, had called this a strange sign. And indeed it was. But Tarak had seen many other strange signs in the dirt – signs that, for now, he would keep to himself.

The clan gathered inside the cave far from where the sunlight bled in from outside. Dom stood atop a flat rock, waiting for everyone to direct their attention to him.

"We have suffered a great tragedy," he said, attempting to sound sober. Joba was of no spirit to translate so Dom simply repeated everything he said in both Umae and Hek. "A great tragedy at the hands of murderers." He dragged the last word out for as long as he thought practical, drawing emphasis to it. "Your safety, our safety, is my greatest concern. We must deal with these killers."

The clan members muttered to each other, the Hek and the Umae sharing the same basic reactions. Tarak and Nagham stood to one side near the back, listening closely. Tarak seemed to bear a great weight from the sight of the dead. Contemplative, Nagham wanted time to consider all that had transpired. Dom's words were something she wanted to hear now.

"Why?" called a voice from the rear. "Why would anyone do something like that? We have no differences with anyone."

"We do now," said Dom. "Now we do have differences. I don't care why they did what they did. I care only for vengeance."

"Vengeance?" asked Nagham. "Of what value is vengeance? Does vengeance make us safer? Does vengeance serve the clan?"

Dom frowned. Nagham was a nuisance. He had underestimated her intelligence. Now she was questioning him in front of the entire clan. And they revered her. She was, after all, a Gift.

"Yes, it does," replied Dom. "For if we do nothing in response, those responsible will have nothing to prevent them from taking more of our lives. But," he added with another pregnant pause, "..........but, if we act swiftly they will not choose us as the targets of their sickness again. If we are decisive enough, they will choose no one for such a thing."

"Who?" asked Nagham. "Who is it that you speak of? We know of all the clans who roam nearby. We trade with them. We bear young together. Some of our clan members once belonged to those clans. None would do such a thing."

"But they plainly have," noted Dom. "Tell me Nagham, where did you come from?"

The crowd was very interested now. It parted so Nagham could move forward.

"The island. The lore is that I was brought ashore by the island peoples. They left me at the water's edge where I was found by my clansmen."

Dom smiled. "A 'gift', is that right?"

Nagham nodded. "That is the lore. I have lived my life so that others would deem me worthy of such a label."

"It's a lie," said Dom coolly. The crowd broke out in a series of hushed rumbles as the exact translation was shared by those able to do so.

"A lie? What is a lie?" asked Nagham.

"You are not from the island. You were not left by the side of the lake. A group of Hek took you by force from a group of Umae traveling to the Standing Stones. You know of that place?" he asked, looking around.

A powerfully built Hek male stepped forward. "Anshu know. Anshu see. Stones far. Most not see, but Anshu see."

"You have traveled far?" asked Dom slowly.

Anshu nodded. "Anshu take many steps. More than any. Anshu know stones."

"Has anyone else seen them?" asked Dom generally. No one else stepped forward. "No. But I have. I did when I lived on the island before the Great Rain."

"Impossible!" exclaimed Nagham. "You are a child. The Great Rain was too long ago."

"The green sky. The purple lightning. The endless hiss of the water. The floods. Oh yes," said Dom confidently, "I remember. But you have forgotten. I have taken far more steps and have seen many, many more fires than the oldest among you. I am a Journeyer. I am not like you."

"You seem to know much of me," noted Nagham.

"And I know more still." The clan was settling down now, eager to hear the rest of Dom's story. "Some of your group, the group from the island, survived the Heks' attack. They shared with us the story. We decided not to search for you. It was too dangerous." He glanced over at Anshu, appreciating the Hek's powerful build for the benefit of his listeners. Was it so hard for them to see how this could have happened? "But you ended up here, with this clan. Nagham, you said yourself that this group trades with other clans, that you bear young with them. You said that members change from one clan to another. Once the Hek were done with you, they gave you away. Perhaps they got something in return. But here you are nonetheless."

"But the lore says you are wrong!" said Nagham. She looked around at her fellow clan mates. "You have all heard the stories!"

"But does anyone here remember?" asked Dom. "Does anyone remember a time when Nagham was not here?"

The clans folk looked around at each other. It was a young group. They were on the move almost constantly searching for food, water and shelter. Nagham was among the oldest in the cave. No one stepped forward.

"Why? Why make such a lie about me?" demanded Nagham. "Why would the lore say I was a Gift if I was not?"

"You are a clan leader, are you not?" asked Dom.

"Yes!" said Nagham defensively. "Chosen by my people!"

"A gift of leadership?" asked Dom rhetorically. "Tell me, how many of your people died after the Great Rain? Were you able to explain to them what had happened? And now, a number of your clan mates are dead, including Gondiar and Pol. Were they not clan leaders as well?"

Nagham's face glared. "What do you accuse me of ?!?"

"You are angry," said Dom. "Why? Is it because you have been exposed? One clan leader is away and two others are now dead. Tell us, suppose Tarak had been with that group as well. You would be the only clan leader left, isn't that right? Isn't that right, Child of Another Clan?"

"Enough!" shouted Nagham. "You lie. You lie and I will show the others your lies!"

"Oh?"

Nagham gathered her spear and prepared to venture outside. "I will face the Sky Fire now," she declared. "My spirit will walk in the flames and when I return I will show you the truth." The crowd split as she strode towards the cave opening. Tarak went after her.

The clan seemed at a loss as to how best to interpret what had just happened.

"Beware her lies," said Dom quietly. "And be prepared. She will bring death to us if we are not careful. She is gone. Listen to me, my people. I will protect you."

Nagham had already made it beyond the ledge overhanging the cave opening and into a nearby cluster of trees before Tarak caught up with her.

"Nagham, you must stay," implored Tarak.

She stopped underneath the shade of one of the trees. The Sky Fire glared down at them from above. "He lies Tarak. No one here knows, but there are those who know. I will seek them out, those who can remember when I was found. The lore tells the truth."

Tarak nodded. "Dom does not."

Nagham cocked her head. "Speak."

"Tarak listened to the dirt by the dead. It told a strange story. The dead would have walked there or would have been carried by their killers. But the dirt did not tell me that. The dirt told me there was only one person who had been there. That person made small marks on the ground. Marks the size of a child's."

Nagham was processing Tarak's statement. "But how can that be?"

A great weight settled on Tarak's shoulders. "Dom has frightened a daggercat. He has carried Alitus a great distance all by himself. Joba tells me so. He is a gift, but a terrible one. A gift from something dark. There is only one answer as to who the killer was."

Nagham nodded. "We need help."

Tarak was swayed by her conviction. "Yes, but be swift," he conceded. "He must not suspect. I will do as I can to keep the people in line." He placed his hand on her shoulder. "Swift. We must not let this continue for too many fires. Our clan will be lost."

Nagham firmly gripped Tarak's shoulder. "Swift indeed," she replied with a nod. Nagham then turned and dashed off into the woods, avoiding the eye of the Sky Fire whenever she could.

Various members of the clan began returning to the cave shortly before dawn arose in the morningside sky. Some bore game from the night's hunt — rabbits, deer, and even a large pig borne by the hunters on long poles. They settled in with their prey next to a fire just outside the cave to begin skinning and carving the meat with sharp pieces of flint. Others had brought water back from a nearby stream in dry leather skins. They nodded enthusiastically at the bounty gathered by the hunters.

Dom oversaw all this activity with approval. The Umae, the Hek, and their Hybrids operated efficiently together. They all seemed to know their roles without being told. He foresaw no problems in leading this group to a position of dominance.

As the first light crept up on the horizon, a figure appeared in the morning mist. Another figure appeared, and then others, until several dozen were standing nearby watching the cave. The clans folk noticed them but didn't show any signs of alarm. The figures began to move closer causing Dom to rise and focus on their approach. It was a collection of Hek, Umae, and Hybrids. They all bore spears and sober expressions. Nagham was among them. She stood at the front of the center of the group resting the butt of her spear on the ground.

"Your 'proof'?" asked Dom from near the cave entrance. He advanced towards her as he spoke.

"And more," said Nagham. Her voice was calm and sure.

"You seem to have brought a great number of witnesses," noted Dom, glancing about at her companions. "But none of these people appear to have enough ellipses to me." He grinned at her. "You must have brought them for some other purpose."

"Akeed comes as well," said Nagham. "She has more ellipses than any other and her mind is as clear as rainwater."

Dom peered around behind them. "Akeed? I don't see anyone resembling that description," he said. "I can only assume she will be here soon?"

Nagham's eyes narrowed. "She requires aid. Her mind is clear, but her legs are frail. She was a Gift as well."

"Then why keep her at all?" asked Dom. "She doesn't sound very useful."

Nagham's group muttered to themselves. "There is something foul about you," she said finally. "You are a liar. You are a killer. You are a danger to my clan. You will be removed."

"Oh?" said Dom, his face brightening. "By whom? All of these people? Removed to where?"

"I don't care where you go," said Nagham. "But you have accused me of lying to my people. You have accused me of sending others to die for my own benefit. My honor permits me to challenge you and your accusations."

The Hek from the clan were attempting to follow along. Joba gathered them close together so he could translate what Dom and Nagham were saying.

"And mine does the same for yours," said Dom. "What sort of challenge would you like?"

Now it was Nagham's turn to smile. "Our ways would allow me to demand settlement by combat, but I have no wish to slaughter a child. We will await Akeed. She will speak the truth."

"Perhaps you overestimate yourself," said Dom. "But, very well. I will make a bargain with you. When Akeed arrives, we will speak with her. If she supports your side of the story, I will leave voluntarily. But if she doesn't......"

"There is no danger of that," said Nagham confidently. "But if she doesn't, I will withdraw my accusations and agree to you becoming a clan elder. If you can manage that miracle, we would benefit greatly from your leadership."

Those in her group able to understand her comment laughed. As the laughter died off, they turned and saw a litter carried by 2 Hek and 2 Umae. Aboard the litter was an Umae woman. She had pure white hair and sat slumped to one side. Dom could see the loose skin around her neck and the deep wrinkles in her face. She leaned forward slightly, trying to study Dom as best she could with her ancient eyes. Her litter bearers helped her stand next to Nagham. She took Akeed's arm and helped her shuffle slowly forward until she faced Dom.

Akeed's eyes were so light they almost appeared to be solid white. She squinted slightly as she stared intensely at Dom for an indeterminate amount of time.

"Well?" said Dom finally. "Don't you have anything to say?" The others gathered nearby strained to hear what was being said. In a quieter voice, Dom added, "You must be tired. Why don't you just move your lips but remain silent?"

His voice nearly toppled her backwards. Her head swam and she felt her balance beginning to waver. After steadying herself on Nagham's shoulder, she attempted to form words with her lips that never matured into sound.

Nagham was wide-eyed. "Akeed?" she asked, a hint of desperation rising in her voice. "Akeed? What are you doing?" Her voice was rising, causing the interest of the others to rise as well. "Tell him!"

Dom smiled. "Nagham," he said in the same low voice. "Come here." Powerless to resist, she leaned down so Dom could whisper in her ear. His low voice flooded her mind, washing away any thoughts she had had before. As she rose again, Dom glanced back at the members of the clan.

"I am a liar!" she called in a dreamy voice. "Akeed has spoken the truth. I am no Gift." She gestured towards Dom. "It is he who is the Gift. It is he you should listen to."

Tarak had been standing next to Joba, listening to his translation. "Nagham?" he called in protest. He lifted his spear and began moving forward.

"And I brought these people here to enslave you!" Nagham screamed at her clan mates. "But I am overwhelmed by my guilt! I seek redemption!" With that she seized her spear and hurled it at the nearest member of Akeed's clan. It impaled a surprised Hek in the chest, toppling him backwards in a rush of crimson.

"These are the killers!" shouted Nagham. "Forward my people! To freedom!" Akeed stood unsteadily, a vacant expression on her face as she continued to mouth empty words.

Both groups immediately broke down in disarray. Tarak's approach was halted by confusion as Nagham struck down the Hek man. In his hesitation, he was struck in the neck by a spear thrown by a member of the opposing clan. Clutching his wound he fell to the ground, his shoulder and upper arm covered in blood.

"Now, my people!" cried Dom. "Follow me!" He pulled the spear from Tarak's neck and charged towards the other clan. The newcomers delayed, the disbelief of seeing a spear-bearing child rushing forward holding them in place. Dom advantaged himself of the situation, attacking with incredible speed. He swiftly impaled a Hek, seizing his victim's spear as the man fell to the ground. He used that spear in turn to slay the man standing next to him before any of the other clan members could react.

Tarak's death animated his clan. They seized the weapons that were set aside by the hunters and entered the fray. A small cloud of spears soared through the air with perhaps half finding a target. The other clan responded with spears of their own. A group of young adults behind the front line found angles to unleash stones from their slings. Tarak's clan sought refuge near the cave's entrance. Nagham found a spear on the ground next to a fallen Umae, but her reactions were sluggish. As she bent over to retrieve it, a huge Hek male seized her by her hair and rammed her head into the stony ground. She collapsed face first, remaining motionless.

Dom's agility was more than the other clan could counter. He whirled through his opponents, stabbing and battering anyone in his way. Once the barrage of sling stones abated the rest of Tarak's clan moved forward from inside the cave. They were fully armed now. Some pelted their enemies with sling stones while others charged forward, skewering their foes with spears. The other clan broke and attempted to run. Dom circled behind them, cutting off their main path of retreat. As they attempted to run by, he managed to seize several individuals - pulling them down and snapping their necks. Others were slain by another hail of spears and sling stones. Only a handful managed to escape the scene and flee off into the early light.

"Hold!" called Dom. He was exuberant. His face was flush with the thrill of slaughter. "Hold, my clan!" He turned back to them and raised his hands. The other clan members stopped and turned towards him. "We are victorious!" he proclaimed. "I told you that I would lead you to greatness. These villains came intending to slay us all, but we have defeated them." He gestured towards the glow in the morningside sky. "And once the Sky Fire sleeps again, we will find them. We will teach them the price for such intentions, and I will show you the fruits of your victory."

A couple of the clan members examined Nagham. Her skull had been crushed and her face was almost unrecognizable. Tarak's blood covered the ground beneath him. But they had Dom. Dom, who had defended them from death. Dom, who had promised to show them the way to greatness. And they were heartened.

CHAPTER 11

Sol and Lin began studying the data Jack had gathered about Haven as soon as it became available. Fortunately for them they didn't need a Tech Agent to help them understand the portions relating to the biodiversity on the planet. They simply engaged the ship's version of sim-Eve to help them with data analysis.

As was his habit, Sol attempted to use this as yet another opportunity to tutor his student. "What can you tell me about these patterns?" he asked, noting a series of charts that were on the display.

Lin leaned in. "It looks like a typical evolutionary pattern in the various proteins schematics."

"And what about these?" asked Sol, changing the depiction in front of them.

Lin was momentarily lost in thought. "The same," she said finally. "There are a lot of recessive traits that are likely not advantageous from an evolutionary standpoint anymore. And there are still schematics for the creation of vestigial structures within a number of organisms."

Sol nodded. "Excellent! You have become a fine Life Agent." His smile lifted her and brought a warm glow to her cheeks.

"I don't know about that," she said sheepishly, "but you have taught me so much."

"What other information do we need to make more detailed judgments concerning the evolutionary patterns on Haven?"

Lin was feeling pressured to answer every question perfectly.

"We don't have a lot of information on extinct species. There may be remnant schematics, but likely not enough for us to draw any helpful conclusions. We would need climate data and data concerning the planet's geology. Anything from fossils to tree ring studies to rock samples from within the ground that can yield information about atmospheric conditions at any given time."

"Yes," sniffed Sol. "And how would we get that data?"

Lin didn't like the only answer she could come up with. "A Tech Agent?" she asked hesitantly.

Sol nodded. "Unfortunately. But I wonder........" He rolled over to the Life Lab's Alpha panel. Lin followed close behind. Sol pressed his palm against the screen. "I need to access sim-Jack," he said confidently.

The display flickered briefly before responding. "You need a lot more than that," replied the panel.

Sol glanced at Lin and raised his eyebrows.

"Jack, we are trying to understand the data package that you sent us but we need some help. None of us are Technical Agents and some of the science is beyond us."

"You have a talent for noting the obvious," said sim-Jack.

"Can you help?"

"It's not a question of 'can'. It's a question of 'why do I care?'"

"I just want to make completely certain that Haven becomes the environmental paradise that Jack wants it to be. He's going to be the most famous Umae ever and I can't think of anyone more deserving."

The display remained silent until Sol and Lin began to wonder if sim-Jack was even still activated. Just as Sol moved to check, sim-Jack responded. "Be brief. What do you want?"

"We are trying to assemble an environmental history. We need information about core ice gas levels, tree ring growth, a profile of elemental concentrations in the planet's crust, and an analysis of Haven's sun. Is there anything unusual about its solar cycles? How old is it? That sort of information. I'm sure you will be able to think of additional data that will be helpful to us."

"No doubt," grumbled sim-Jack. "Very well, since all of this information is already available, I'll dumb it down for you. There. You can access it on the panel where you were working previously. But I'm not going to explain any of it. If there is something you can't figure out, you will have to ask that idiot Mac."

"We will do our best," said Sol. "Thank you. We couldn't do this without you."

"I know." With that, sim-Jack disengaged.

Sol and Lin returned to the prior panel. As sim-Jack had mentioned, the data was waiting for them.

Sol again placed his palm on the display. "I need to access sim-Eve."

"Hello, Sol." The Eve simulation was plainly much more pleasant to interact with than was sim-Jack.

"Eve, I need for you to analyze the data on the display and try to determine any patterns or trends relating to atmospheric gasses, sun cycles, and the chemical composition of the subsurface rock. How long will that take?"

"What level of detail do you require?"

"As much detail as you can give me."

The display swirled with color briefly. "Come back in a quarter of a rotation."

Sol nodded. "Very well." He turned towards Lin. "That will give us time to study the micro-organism profile of the planet. I fear Jack has underestimated that threat. Since he didn't share his methodology, I'm going to have to check his work."

"Won't that make him angry?"

"I doubt he will notice. After all, he is locked up in his room and his sim doesn't pay any attention to morons like us."

Lin stifled a laugh. "You managed to get sim-Jack to do more for us than I would have expected."

"That was a part of your lesson on DSM," explained Sol. "You can't interact with those afflicted like you do with everyone else. Appeal to their sense of superiority, to their sense of uniqueness. You will be surprised by what you can get out of them."

Lin nodded. "I'll try to remember that. Now, let's check out those germs."

Lin threw herself into the analysis of Jack's Haven data. She had also redoubled her studies of DSM, mindful of Sol's opinion about the likelihood of her responsibilities with Jack. The combination pushed her endurance more than she had anticipated. She found herself ingesting stimulant concoctions she prepared in the Life Lab outside of Sol's observations just so she could keep up. Perhaps she could rest when all of this was over.

Sim-Eve found all sorts of different trends and correlations in the Haven data. But finding the trends and correlations was one thing. For the data relating to geology, chemistry and the physics of Haven's sun, she needed a Technical Agent to explain the reasons behind the trends and correlations. Otherwise, the charts and graphs meant very little to her. Given sim-Jack's increasing hostility, Mac was her only option.

Mac's moods had brightened considerably ever since Jack had started spending all of his time in his secret chamber. Mac had stopped using sim-Jack entirely and was determined to figure things out strictly through the use of the tutorials and his own intellect. Lin moved up behind him as he was lost in his studies on the *Aurora*'s Alpha panel causing him to startle when she spoke his name.

"I'm sorry," said Lin trying not to laugh. "What are you so invested in?"

Mac took a deep breath. "Everything," he said. "I really feel like a lot of this material is starting to come together."

"Great. And you did that without any help from Jack or his sim at all."

"Despite their help in some cases," noted Mac. "Did you need something?"

"Yes. Sol and I have been analyzing the Haven data. There is information about a broad variety of the planet's characteristics. Sim-Eve has provided us with some analysis concerning its geology, atmosphere, and its sun's solar history. We need your help in understanding what that data can tell us."

Mac quickly rose from his chair. "Certainly. Is Sol in the Life Lab?"

"Yes," said Lin as she started in that direction. "He sent me down here to get you."

Mac hesitated. "Really? He asked for me?"

Lin paused as well. "Why are you surprised by that?"

Mac shrugged. "It's just that Jack has been talking about what an idiot I am every chance he has gotten. How do you know he isn't right?"

Lin considered what she had learned from her studies of DSM. "He thinks we are all idiots," she said. "But Sol isn't an idiot. And I'm not an idiot. So it only stands to reason that you aren't an idiot either. Neither of us think that."

The young man loosed a long sigh. "I hope you are right."

Lin continued walking towards the Life Lab. "Show me," she said warmly.

The walk to the lab was a short one. As they entered, they noted that Sol was slumped over in his chair. Lin quickly raised a hand to counter Mac's concerned expression. "Is he all right?" Mac asked quietly.

Lin walked over next to Sol and gently held her fingers to his wrist. "Asleep," she whispered. "He sleeps a lot now. Not all that unusual."

"Is he ill?"

Mac's question nearly paralyzed Lin. Oddly, it was one she hadn't anticipated. She was technically Sol's Life Agent and the old man had an expectation that she would keep his health condition confidential. But his health was also a central consideration of the entire mission. At some point, the other crew members would need to know.

"He has a lot of ellipses," she said vaguely. "Now, tell me about this data."

She guided him over to a nearby panel and activated the display. A series of graphs and charts awaited his review. The first involved readings from Haven's sun.

"Pretty basic stuff," said Mac as he looked at the screen. "Haven's sun is about 20% older than Uma's sun, but they are very similar. Haven is also very nearly the same distance from its sun as Uma is from its sun. The solar cycles are very similar. I'm not seeing anything about this sun that would cause it to affect Haven any differently, statistically speaking, than Uma's sun affects it."

"From our collective knowledge of hard sciences, that was sort of what we were guessing too," said Lin. "But these next studies were more confusing." She brought up the next set of data.

"Geology," noted Mac. He peered at the tables and charts in front of him. "But I'm not seeing anything that should have confused you."

"Just wait," said Lin as she brought up the next screen.

"Polar ice studies," said Mac quietly as he began his review. "Wait. Is this data accurate?"

Lin shrugged. "Jack collected it. We have no reason to think that it isn't."

"And I've reviewed the *Aurora*'s sensor arrays," said Mac. "They are remarkably accurate."

"So what are you seeing?"

"Well, the gasses trapped in the polar ice at various depths strongly suggest a wide variance in Haven's temperature history. For a long time, probably hundreds of thousands of ellipses, its temperatures weren't extraordinary. There were typical fluctuations caused by sea current changes, sun cycles and the like, but all of that is normal. But look here." He highlighted a portion of the graph on the screen.

"It spiked," noted Lin. "We saw that. But why?"

"Now I understand why you had me look at the geological data first," said Mac. "Because it is perfectly normal. So we can't attribute the temperature spike to volcanic activity or some irregularity in the planet's core. But we also can't attribute it to any sort of insulating effect."

"What do you mean?"

"A planet's temperature can rise suddenly, relatively speaking, if certain gasses build up in its atmosphere for some reason. Those gasses insulate the planet and trap heat that would otherwise radiate out into space. But the gasses in the ice cores aren't the right sorts of gasses to do that. In fact........" He took another long look at the ice core data. "I'd say the opposite happened."

"The opposite?" Lin was clearly confused and Mac was actually enjoying the process of explaining all of this to her.

"These gasses indicate a sharp jump in radiant energy. Sunlight. The spike is so abrupt that I'd say some dramatic event took place that significantly thinned out the atmosphere. More sunlight was able to reach the planet and the temperature rose quickly over a very short period of time."

Lin frowned. "That matches the studies we had on fossils. The evolutionary patterns followed by Haven's lifeforms placed a premium on the ability to survive in hotter climates with more direct sunshine. The atmospheric change was so fast that it made survival very difficult for most species. Even now, the number of species we would expect to see currently based on the fossil record is rather small. A lot of species couldn't handle the climate change so they became extinct. So here's the big question – what caused it?"

Mac closed his eyes in thought. "Creating a thicker atmosphere is relatively simple," he said. "Volcanism. A giant meteor. Both of those can do it. But thinning one out? That's a bit trickier."

"Any guesses?"

"The most likely way would be to reduce the amount of water vapor in the atmosphere. That could cause a significant thinning of the cloud cover which would result in a temperature increase."

"A big rain storm?"

"An extremely big rainstorm. If she didn't already, you might have sim-Eve check the geological data on rock strata that appears to be relatively unaffected by its submersion. Check for erosion, interactions between rock formations and salt, that kind of thing."

"But what would this......unaffected rock tell us?"

"It would tell us that the submersion was sudden and relatively recent. You see, eventually a planet's water cycle will even itself out. If something is done to throw off the natural water-water vapor ratio, the ratio will eventually re-establish itself. But it takes a long time. Haven still has a relatively thin atmosphere, at least compared to Uma's. So if something happened to transform a lot of its atmospheric water vapor into rain, it would have had to have happened fairly recently. At least in terms of geological time. There would have been a great deal of flooding with new rock strata becoming submerged. That rock would not have been submerged for very long, so it would appear to be relatively unaffected by the water. If you find that, we've found our answer."

Lin added some directives to the panel asking sim-Eve to begin the process of making the analysis Mac described. "Next big question – what caused that?"

"That's where it gets interesting. It would require a lot of energy. The energy would have to be generated by an atomic reaction. If that much energy were directed at the planet through some natural means, like a solar flare of a stray gamma burst, there would be other indications of it. But there aren't." He paused to make sure he was comfortable with his conclusion. "That means it was directed by some intelligence. An intelligence advanced enough to execute an energy pulse of that magnitude without doing any noticeable collateral damage."

Before Lin could reply, the panel indicated it had completed the analysis she had requested.

"There are numerous areas of exposed rock stratum that are submerged in saltwater," said sim-Eve. "Over 90% of those strata are substantially the same as stratum of the same type of rock that is exposed to open air. They are essentially unaffected by submersion."

Lin and Mac looked at each other in silence. "An intelligence?" asked Lin finally.

Mac nodded. "I don't see any other answer."

"So, third big question – why would an intelligence do that?"

Mac shrugged. "Not my area of expertise. I'll let you and Sol figure that one out." He looked over at the old man who was still asleep in his chair.

"I'll talk to him when he wakes up," said Lin. "And Mac? You are far from a moron. While my heart wants you to be wrong, my brain is telling me that you aren't. But it looks as if some alien lifeform took steps to make our new home uninhabitable for us. They failed, but that doesn't make me feel any better. And what if they are still out there?"

Mac stood up to leave. "Maybe they didn't know we were coming," he said, trying to sound optimistic. "But if they are still out there, they've been 'out there' for a long time. I just hope our technology can find them before theirs finds us. Because, one way or another, we are going to have to land on Haven."

Sim-Min brought the **Starshine** into orbit around the faint blue planet. Wispy white clouds floated below them offering glimpses of the planet's surface. Sim-Min had requested sim-Eve to initiate a new data study.

"Sol, are you able to confirm the conclusions you reached previously concerning the biodiversity and its potential threats to us?"

Lise looked up from her own display, waiting for sim-Sol's reply to sim-Eve's inquiry.

"The diversity is rather narrow, suggesting early stages of evolution from more primitive common ancestors," noted sim-Sol. "But that makes the analysis easier. There are a few large land animals, both carnivores and omnivores, that could potentially view an Umae as prey. There are a number of smaller species who could pose a threat if disturbed unexpectedly. They would not be expected to initiate a confrontation."

"What about microorganisms?"

"The range of microorganisms is similarly narrow," replied sim-Sol. "The number of species is surprisingly low. It would be what I would expect following some sort of planet-wide cataclysm."

Lise swallowed. "Maybe not the best choice of words."

"The number and distribution of species fits a pattern predicted by a sudden decrease in life forms generally. A few species were already adapted to whatever killed everything else off and survived. But their mutation rate has not permitted them to evolve into very many new species yet."

"Sol," said Lise, "based on that can you estimate when such an event might have taken place?"

Sim-Sol remained silent for a moment. "No, I can't. I don't have enough information about the nature of the organisms that existed before such an event. That precludes any meaningful calculations."

"Are there any Umae down there?" asked Lise anxiously.

"The sensors have not confirmed any life forms consistent with the Umae," reported sim-Sol.

"But they should have gotten here already," said Lise.

Sim-Min interjected before Lise's heart could sink completely. "There are two areas where the sensors were unable to gather any data at all."

"How can that be?" asked Lise. "Where are they?"

"There is a relatively large structure near the planet's equator. Its exterior is somehow resistant to our scans. I could increase the energy of our sweeps, but it might endanger anything living inside the structure."

"This structure," continued Lise, "what is it?"

"It is a four-sided pyramid, larger than the Chroniclers' Hall on Uma by a factor of ten."

Lise sat back in her chair attempting to extrapolate from what she just heard. "Another pyramid. So, there's a connection between Uma and Haven."

"It would appear so," concurred sim-Min.

"What about the other structure?" asked Lise. Her voice was shaken by the first hint of panic.

"It's a Jouneycraft, almost certainly the *Aurora*."

"The *Aurora*?" echoed Lise. "Are Journeycraft scan-resistant?"

"No. But I can only tell you what the scans indicate," said sim-Min. "While the scans can't penetrate its exterior, its dimensions are exactly those of the *Aurora*, apart from what are likely minor modifications made by Jack."

"They could still be onboard," noted sim-Sol.

"Or they could be inside the pyramid," added Lise. "Let's think this through. We don't know how long ago they got here. But Jack would have had sensor data very similar to ours, even accounting for any upgrades he may have made. He would have known that indigenous lifeforms didn't present a substantial threat to him or his crew. And unless his sensors were able to penetrate the pyramid, his curiosity would have been piqued by the mystery of what, or who, is inside that pyramid. I'm betting he went in."

"If they are still alive," said sim-Sol. His simulation had modulated its voice to soften the impact of his comment, seemingly aware of Lise's anxiety. "The possibility that they are not would also provide an explanation as to why they aren't showing up on the scans."

"Can we look for traces of their presence?" asked Lise.

"I already have," said sim-Sol. "There weren't any. But this planet has wind and rain, so the absence of trace findings doesn't mean much."

"What else is down there, Min?" Lise was easily integrating her skills with those of her simulated comrades. It took her a moment to realize what she had done.

"There are other structures spread about the planet in no discernible pattern that could, with a moderate adjustment in definitional parameters, be called 'Journeycraft'. All of them have significant structural compromises and none emanate any sort of internal energy signatures."

"What does that mean?" asked Lise impatiently.

"They have been sitting there for a long time. Their external structures have broken down or they have been damaged. The lack of energy signatures would likely indicate that their engines are either non-functional or have been removed."

"How many?"

"Sixteen."

"Wait," said Lise. "There are sixteen broken down Journeycraft on this planet's surface? Are they from Uma?" A cold wave rolled down the back of Lise's neck.

"Impossible to tell. Interestingly the remnants of the Journeycrafts are very similar to a degree suggesting they were initially identical."

"Is that important?"

"Possibly. Every Journeycraft's Command Agent was free to improvise and add modifications to his or her own ship. The differences between the *Starshine* and *Aurora* in terms of their technical capacities were marked. These ships are almost identical in every way."

"So you are saying that these ships are not the ones from Uma that have seemingly disappeared?" asked Lise.

"That's highly unlikely. If that were the case, we would see much more variation in these ships."

"But if these Journeycraft didn't come from Uma, where'd they come from?" asked Lise.

"An excellent question. But there is more. There are twenty-nine smaller structures that are also spread around the planet's surface that more clearly fit parameters we are familiar with."

"How so?" asked Lise. "What are they?"

"Based on their size, composition, and the nature of their structural compromises, it is a near certainty that they are beacon satellites."

"On the surface?"

"They all fall within the predicted profile of a beacon satellite that has lost power and fallen to the surface."

"Eve," began Lise unsteadily, "I think you understand how much I'd like to see Mac again. So forgive me if my logic here is flawed. We have a lot of questions that need answers, and we have a lot of data from everywhere on Haven except for one place."

"The pyramid," replied sim-Eve. "Your logic seems sound to me," said Eve. "Sol, would we need to take any environmental precautions for the planet's surface?"

"None, aside from a good supply of water. The average temperature is approximately 32% warmer than that on Uma."

Lise stood up. "I'll fill up some canteens," she said. "Min.....sim-Min?"

"Yes?" replied the simulation.

"I realize that you are still in charge, but can you take us down? I want the rest of those answers and that's where we will find them."

"That was going to be my recommendation," said sim-Min. "Well considered."

Lise unnecessarily braced for the landing. As it was her first one, she had expected to experience a stronger impact. Instead, sim-Min had guided the **Starshine** to Haven's surface, setting it down with a touch so feather light that Lise felt almost nothing. She remained seated briefly to make sure the ship wasn't moving before rising from her seat and walking to the egress portal. She had decided that she wasn't going to wear an environmental suit. According to all of the scans and calculations done by all of the sim panel personalities, there wasn't anything dangerous on the surface. The microbial life was benign to Umae and the proximity scanners had confirmed that no animals of any size were close enough to pose a threat to her. What she really wanted was to see Mac and to be able to hold him as soon as she did. A vacuum suit would just get in the way. But it had been a long time. She hoped that he felt the same way.

She walked down the ramp to the surface. Clumps of tan grass were scattered about on the ground and she could feel a warm breeze on her face. The air smelled slightly sharper than what she had remembered from Uma. The dirt beneath her feet was a rusty brown. She could see a number of scrub trees growing nearby. A few of them held blossoms amongst their leaves. Just off to her emptyside was a Journeycraft. It looked very similar to the **Starshine**. The sims had opined that it was almost certainly the **Aurora**. Mac and the others had made it at least this far.

Beyond the **Aurora** towered the pyramid they had studied from orbit. The specifications provided by the scanners concerning its size didn't do it justice. It was enormous. She tilted her head way back as she searched upwards for its apex. Just like the Chroniclers' Hall on Uma a large antenna extended from its peak. But with whom had that antennae communicated?

She glanced back up the ramp. Eve was still recovering from her interface with the enormous beacon satellite. She had given Lise such a grave warning about avoiding this planet. But the **Starshine's** scanners hadn't detected any danger at all. The interior of the pyramid was a mystery beyond the reach of those scanners. Lise didn't know what to expect. The source of Eve's apprehension, the other Journeyers.......either or both could be inside. The mechs couldn't remain in communication with the ship once they entered the pyramid. If something happened to prevent them from coming back outside, Lise and the sims on the **Starshine** would have no way to tell what it was. Eventually, an Umae would have to go inside. Since Eve was unavailable, Lise was the only option left. Despite the warmth of the wind, the sight of the other ship and of the pyramid caused a shiver to rise up the back of her neck.

Lise activated the com-link next to her mouth.

"Min, I'm going to take a look at the **Aurora** first," she said. "Are you reading me?"

"Yes, I can read you." The sound of his voice brought her a measure of comfort.

"I'll let you know when I get there."

"You will want to stand near the position where the egress ramp would drop," said sim-Min. "I'll give you the instructions for access then."

Lise nodded, forgetting that Min, sim or otherwise, wasn't actually with her. Even her pace towards the ship was conflicted. She desperately wanted to know if Mac was alive, but she was deathly afraid that he was not.

"I'm here." She looked up at the belly of the **Aurora**, trying to determine where its ramp might settle. The ship's landing piers held it aloft allowing her to stand beneath it. She knelt down and studied the ground hoping to find some sign of passage. She saw none and was reminded of sim-Sol's comment about how the weather patterns experienced by the planet would likely have eradicated any secondary evidence of the crew. If Mac's footprints had been here, they had been erased by the wind and the rain.

"There should be an access panel on one of the port landing piers," said sim-Min. "Find it and tell me what it looks like."

Lise looked up and down the landing piers until she saw a collection of buttons about shoulder high. "Found it," she reported. "It's a collection of buttons and ports. Sixteen buttons arranged in a four by four pattern. There are...... eight ports in a straight horizontal line. They are all the same size."

"Take the device you brought with you and plug it into ports three, four, seven and eight," said sim-Min. "Let me know when you have done that."

Sim-Min had directed one of the mechs from the *Starshine* to build the device she now attached to the access panel. The insertion tabs adjusted themselves to insure they were the correct size for the ports. "Done."

"Now I will interface with the *Aurora* using my Command Authority to......." His voice was interrupted by a screeching howl of static.

"Min?!?" called Lise frantically. "What's wrong?"

The screeching continued for another moment. "Remove.......device......"

Lise seized the device she had been given and pulled it away from the access portal, breaking the connection. "Min?"

A brief echoing pulse rang through her earpiece before falling silent.

"Odd," said sim-Min. "It appears that Jack booby trapped the access panel."

"Booby trapped?"

"Yes. As soon as I gained access, a wave of malignant electronic protocols attempted to flood my system. I have managed to quarantine them. My initial analysis indicates they were intended to deactivate the *Starshine*."

"Deactivate? But why would Jack want to do that?"

"It may not have been Jack's intent," replied sim-Min. "The protocols involved were designed to attack practically any electronic system. The fact that it attacked the *Starshine* could have been a coincidence. They may have been a generic defense mechanism Jack devised. I will need more time to analyze them."

"But we can't get inside?"

"No. Any additional efforts would likely result in a similar attack. There is no guarantee that I would be able to successfully defend the *Starshine* again. I would recommend against it."

Lise turned and faced the pyramid. "There is really only one more place to look."

"Yes," concurred sim-Min. "You can still eliminate one of the two places that we have been unable to scan. If they aren't in the pyramid, then they must be aboard the *Aurora*. Or....."

"Or?"

"Or they are dead."

Lise was not quite ready to leave the Journeycraft sim-Min had almost definitively identified as the *Aurora*. The fact that there had been electronic defenses protecting the access panel buoyed her confidence that the crew was still alive. The gigantic pyramid in the distance opened a chasm of awe deep in her gut that made her extremely uncomfortable. Her hope of finding the others on board this Journeycraft joined with her apprehension in approaching the pyramid to create an incentive for her to find some other way to get inside the ship. But if the access panel wasn't the answer.......

She leaned down and picked up a dark gray rock just larger than her hand and began striking one of the landing piers. The clang of rock on metal interrupted what had otherwise been a calm, quiet rotation. After she had struck the pier several times, she began to wonder how far away the sound could be heard.

Sim-Sol had advised her that there weren't any large animals in the vicinity. Even so, there were enough scrub trees, prairie grasses and rocks strewn about the general area that she wasn't confident she would detect a threat before it was too late. Certainly she could expect a timely warning from sim-Min, but her nerves would be more settled if she could assure herself of being able to detect danger without that type of assistance. She glanced at the rock before tossing it aside. There hadn't been any response that she could discern from the ship. Her solitary option loomed before her, its apex reaching high into Haven's light blue sky.

There wasn't anything indicative of passage by the other crew along the way to the pyramid. Although sim-Sol had told her that would be highly unlikely given the dynamic nature of the planet, she had held on to hopes of seeing something anyway. The walk to the pyramid took her much longer than she had estimated, additional evidence of its massive size.

She hesitantly reached out and placed her palm on the stone. She could barely make out slender seams in its craftsmanship. Each component stone around the base was at least five times as high as she stood. Its surface was almost completely smooth, despite what sim-Sol had said about the changing weather patterns. She walked all along the base of the side that faced the Journeycraft she had visited earlier. The stone was completely uniform. There weren't any panels or doors.

"Min, can you hear me?"

The prompt response from the **Starshine** eased her nerves slightly. "Yes, Lise. You are at the pyramid. I've been attempting to determine how you might get inside."

"Good, because I have no idea. This stone is unlike anything I've ever seen. The craftsmanship is comparable to the Chroniclers' Hall on Uma, but the stone is extremely smooth. Hard to believe it has been out here in the wind and rain for so long."

"The conclusion that the pyramid is relatively old is based on circumstantial evidence," noted sim-Min. "Scans are unavailable, even of the material comprising the structure. But it is very likely that the builders used some sort of treatment on the stones to make them more resilient. The types of rock that Haven itself is composed of do not share that type of durability. It's more reasonable to conclude that the stones were bolstered chemically than it is that the stone was brought here from some other body. Continue around the perimeter of the structure. If there is a means of ingress that is set into the wall of the pyramid itself, it may have evaded my scans. But it may be obvious enough for you to see it."

"I'll do it," said Lise. She continued along the base until she reached the corner, moving slowly enough to allow her to study the stone face as she walked. The exterior appeared to be perfectly uniform. The seams looked the same and she couldn't find so much as a scratch on the surface. "Continuing along the morningside perimeter," she advised sim-Min. "Threat level still minimal?"

"We'd tell you if that changed," said sim-Min. Lise could imagine an Umae responding to her question in a tone dripping with annoyance. Sim-Min either had no such capacity or chose not to employ it.

The morningside perimeter looked exactly like the one she had just studied. Again, she made it all the way to the next corner without seeing anything noteworthy. "Moving to heartside," she said.

"Lise, stop." It was sim-Sol. Atypically, the voice held a hint of urgency. Lise froze.

"What is it?"

"I briefly detected movement directly ahead of you. It was as if something quickly emerged from the face of the pyramid before returning inside. The pyramid itself prevents us from achieving perfect scan data, so I can't tell you with any certainty what it was. Lise, is the wind blowing?"

"The wind?" Lise hadn't been paying attention. She stood still for a moment and focused on the surface of her skin. "No, I don't really feel anything," she said finally. "Why do you ask?"

"Because if the wind is calm, any motion that we can detect must be attributable to biological energies," said sim-Sol.

"I thought you said there wasn't any threat in the immediate area of a biological nature?" Lise was beginning to perspire.

"Within the bounds of our ability to scan," said sim-Sol. "I assumed you understood that. If something is alive within the pyramid, we would have no way to determine that."

Lise drew a deep breath. Obviously. The sims didn't point that out to her because they had concluded she was a lot smarter than she actually was.

"Do you want to come back?" It was sim-Min.

"Our ability to scan the pyramid isn't going to change, is it?"

"Highly unlikely," said sim-Min. "At least not within several Umae lifetimes."

"And the Umae on Uma may not have many Umae lifetimes," said Lise. "So I'll have to determine what's in there the old-fashioned way. Sim-Sol, how far down was the motion detected? I didn't see anything."

"Approximately seventy-nine point six six eight percent of the way to the next corner," said sim-Sol.

"That's approximate?" queried Lise with a hint of sarcasm.

"A more precise measurement wouldn't be beneficial," said sim-Min. Lise wondered if the real Min would have seen the humor in the simulation's comment. She tracked the base of the pyramid. She couldn't see anything moving now.

"All right. Here goes."

She slowly crept along the pyramid's heartside exterior. She still wasn't seeing anything noteworthy about its surface and whatever sim-Sol had detected hadn't made another appearance. She pressed forward trying to keep her footfalls as close to the base as she could.

As she approached what she estimated to be near the center point of this face, she could see something on the ground up ahead. She didn't trust her ability to judge the size of the object given the enormity of the pyramid she was next to. It was roughly square and not very tall. She waited a moment to see if anything was going to emerge from the pyramid before proceeding. The pyramid's face still didn't give any indication as to how anyone, or anything, could get in or out. Lise leaned in against the pyramid's surface and took small side steps towards the object. As she drew closer, she could see that it was a container of some sort. The lid was slightly ajar. On the adjacent exterior wall was an opening in the pyramid's face.

"Min?"

She pressed her com link to her ear but couldn't hear anything.

"Sol?"

Again, no reply. She tried to look inside the opening. Some type of artificial illumination revealed a short passage that opened into some larger space.

"Min? Sol?" The pyramid was now directly between her and the **Starshine**. Her best guess was that it had now effectively prevented communications in the same way it had resisted the scans. She was on her own now. She stepped inside the opening and crept down the passage. It intersected with another perpendicular passageway. Off to her emptyside she could hear a voice. Carefully peeking around the corner, she saw a figure kneeling down next to a number of containers like the one she had seen outside. It was an Umae and it had its back to her.

Lise's heart was rattling in her chest. She drew back to consider her options. The figure was muttering to....someone. Or maybe to itself. She couldn't quite make out any words.

Composing herself, she stepped back around the corner into the passage. When that didn't draw the figure's attention, she cleared her throat. "Um. Hello?" She said quietly.

"I've told you.....," started the figure as it rose and turned towards her. It was a bearded man with a few more ellipses than her. When he saw her his mouth fell open and he stopped talking. Lise could feel the warmth of her tears well up in her eyes and flow down her face.

"M-mac?"

"Lise!"

She sprinted forward to him and threw her arms around his neck. He was temporarily frozen by indecisiveness until he leaned into her slightly and caught the scent of her hair. The two drew one another into a tight embrace as Lise began to sob.

"I thought....thought you were gone." She said finally. "How?" She stepped back and looked at his beard.

"I just stopped shaving it," said Mac.

Lise giggled. "N-no! It's fine. Really. But, how are you here?"

"Come with me," he said as he took her hand. "You will not believe what we have found."

CHAPTER 12

Lise soon discovered that the interior of the pyramid didn't look like a building at all. It resembled a cave with twisting passageways, uneven floors, and stone walls slick with condensation. Mac led her by the hand down several passages until they reached a gigantic central chamber. It was completely illuminated although the source of light wasn't apparent. There were machines somewhat resembling the panels from the Journeycraft spread throughout the chamber. Most of them were inert but Lise could see Lin intently studying the figures on a panel in front of her. Next to her was a table bearing a figure covered by a metallic blanket. Lise looked at Mac and awaited an explanation. Lin looked up and, upon seeing her, leapt from her seat and ran towards her.

"Lise! Lise! Oh my stars! You're here!" She threw her arms around Lise who happily returned her embrace. Lin stepped back. "You look great. The extra ellipses have served you well."

"And you as well," said Lise. "But…."

"Where is Eve?" interrupted Lin.

"She's still on the **Starshine**," said Lise. "In repose. It's a bit of a long story, but she is going to be alright."

Lin's eyes widened. "Is she ill? Injured? I'd like to examine her."

Lise nodded. "Yes, of course. But sim-Sol told me she is going to be fine. She interfaced with the beacon satellite we found and her processing period in the repose chamber is taking longer than anticipated. But she told me…… Is everyone from your crew all right?" Lise's eyes drifted over to the figure on the table. A dark shadow crept up in her gut. "Sol?"

Lin glanced momentarily at Mac. "Sol is in repose as well," said Lin.

"Is he ill?" asked Lise.

Lin fretted over her reply. "He was growing more fatigued every rotation. It was his decision."

Lise was confused. "Is that…. Jack?" she asked, pointing at the table.

"Oh. No. Not Jack," said Lise uneasily. "Jack is on the **Aurora** working on……something. Come. Let me show you something."

Lin led Lise over to the table next to the panel where she had been working. Lin slowly drew back the metallic sheet revealing the remains of a creature unknown to Lise. It was humanoid and had extraordinarily long arms and a disproportionately large head. Its head was encircled by some type of bony structure.

"What is that?" asked Lise finally.

"We found it in the middle of this chamber when we arrived," said Mac. "There was a second one that we have placed into storage on the **Aurora**. We don't know for sure what they are, but Lin has been trying to work that out."

"Sol did a rather cursory examination of their protein schematics before telling us they were the key to our RTS research. But his energy was fading very quickly. Before he could explain anything in detail, he asked to be placed in repose." Lin's explanation pressed heavily on her shoulders.

"What's....wrong?" asked Lise nervously.

Mac saw that Lin wasn't prepared to speak any more about Sol. "Sol has an advanced case of RTS. He essentially entrusted us to figure out what it was about these schematics that applied to the RTS research. He said he wanted to be there when we found Min, just in case Min is still alive. There is a possibility that both of them could be cured."

Lise smiled at Lin. "He trusts you, Lin. With his life. He wouldn't have done that if he didn't believe in you."

Lin nodded meekly. "We couldn't have done any of this without Mac. He got all of this online for us."

Lise smiled. "You were able to figure out this technology so quickly?"

Mac shrugged. "It was actually very simple. Lise, these panels are very, very similar to the ones we use on the Journeycraft. Too similar for them to just be here coincidentally. There is some type of connection."

"Do any of the panels have stored data?" asked Lise.

Mac nodded. "Yes, but our efforts to decipher and interpret it haven't been very successful."

"For starters," interjected Lin, "Jack insisted that we only use his algorithms for the project, but they didn't work very well. What we really need is a Data Agent who can interface with the panels and extract the information. That's part of the reason why I wanted to examine Eve. But if her repose has been extended for some reason, it might be a long time before she is up to the task."

Lise turned towards the panel. "I can do it," she said finally.

"What?" said Mac, placing a protective arm around her shoulders. "No. It's too dangerous."

"I'm a Data Agent," said Lise firmly. "And if Eve can't do it, then I'm the best Data Agent we have."

"We have taken a lot of readings on this planet," protested Mac. "It is habitable. It IS Haven. I don't see why we need the data in the panels at all."

"Because this place is dangerous," said Lise. "Sim-Sol had me do a personal interface with Eve while she was in repose. Eve couldn't share with me all of the data she drew from the beacon satellite, but she did tell me to send you all a message. She told me to tell you not to land here because it was too dangerous. So unless you know what she was talking about, we need to know what information is stored in that panel."

"We didn't get any messages," said Mac. "And I'd assume we had our com-array open."

"You don't know?" asked Lise.

"Jack had it connected to his chamber," said Lin. "It wouldn't surprise me if he got the message and just ignored it."

"We will still need to get him over here," said Lise. "We will have to get his cooperation on this data transfer. For one, we don't want him interfering. And for two, as much as we all hate it, we will probably need his technical prowess." She leaned into Mac. "Don't take that the wrong way."

Mac pulled her closer. "You are right, at least about the first one," he noted.

"Can you ask him to come?" asked Lise.

"Yes, of course," said Mac. "He has mostly been ignoring us, but news of your arrival should motivate him."

"Good," said Lise. "Lin can help me get ready for the data transfer. After you contact Jack, I'll need you to prep the panel for the interface. I do need your technical prowess for that," she added with a smirk.

Mac blushed slightly as he looked away. "It's nice to be needed. I'll be right back." He turned and headed off to contact Jack leaving Lise and Lin to prepare for the process of learning about what information had been stored on these panels.

Jack's face was expressionless. An outsider would have been unable to tell that he hadn't seen Lise for over five centuries or that he had any interest at all in the fact that the **Starshine** had found him and his crew. He walked in, trailed by Mac, and sat down at a long table. Sitting back, he folded his arms and waited for the others.

Gathering his hint, Lise, Lin, and Mac joined him at the table. Lise's eyes narrowed.

"The aliens had tables?"

Mac shook his head. "No. I made it. Along with the other furnishings in here. When we first arrived, the panels and displays were embedded in the rock walls. We had to extricate them so we could use them more comfortably." He pointed to a number of panels still surrounded by the rocky walls. "They had dozens of panels. We still aren't sure why they needed so many."

"Eve told you this planet is dangerous," interjected Jack. "Tell me exactly what she told you."

Lise's attention was drawn away from Mac's comforting face to the scowl on Jack's that could have frozen a creek. She forced herself to sit up straight before beginning. "She didn't give me a lot of detail," she began. "I don't think she was able to. Her recovery......."

"I didn't ask for your summary!" snapped Jack. "I don't care what YOU think! I want to know EXACTLY what Eve told you. I'll figure out the meaning myself."

Lise didn't allow her stare to slip away from him. "Very well. Let me think for a moment." Lise paused briefly, trying to recall as perfectly as she could Eve's exact message. "All right," she said finally. "She said 'send a message to the *Aurora*. I think Haven could be a trap. Tell them not to land there.'"

"And those are her words and not yours?" asked Jack.

"It was an integrated data exchange between two Data Agents," said Lise coolly. "They weren't exactly 'words'. But the meaning she conveyed is identical to the one I just shared with you."

"A trap," mused Jack. "She was clearly wrong. There was no trap."

"But there must have been some reason for her to say that," offered Mac. "She had just accessed that large beacon satellite, hadn't she?"

Lise nodded. "Yes. And the volume of information she absorbed sent her into an extended repose recovery period from which she still hasn't emerged. She accessed every bit of information that had passed through that satellite for.......who knows how many ellipses."

"Now that's interesting!" said Jack. "Based on the amount of data that had to have been involved, there was obviously a conversation of sorts going on between Uma and whomever, or whatever, was on this end of the tether."

Lise's eyes widened. "But who would be conversing with....them....it.... on Uma?"

Jack scoffed. "Simple. The Chroniclers were the only ones who could access any information in the tether. But as far as sending messages back, I doubt they even realized it was happening. They thought their panels were information storage devices. They didn't realize they were transmitting that information."

Lin leaned forward resting her elbows on the tabletop. "Why would anyone want information sent from Uma to Haven?" she asked. "There isn't anyone here."

"Apparently there was or was going to be," said Jack. "Or these aliens are idiots. I find that possibility to be rather unlikely."

"Maybe that is an answer I can get when I interface with the panel here," said Lise.

"You're doing what?" asked Jack with a laugh. "No, never mind. You're doing no such thing."

"Jack, your data protocols aren't working," said Mac. "The data's format is apparently too different from what we are using to analyze it. We need to use a Data Agent to interface directly."

Jack sneered. "And that's your educated opinion, my dim-witted apprentice?"

Mac stood up. "Yes. Yes it is. Besides, what harm could it do?"

Jack shrugged. "I guess she could end up killing herself or turning herself into a fern," he suggested. He turned back towards Lise. "You go right ahead with your silly attempt at becoming relevant. I seriously doubt you will find anything we haven't found already." He stood up and straightened his shirt. "If you want to take the risk, have at it. Eve will be back on duty eventually anyway so it doesn't really matter what happens to you." He turned and walked off in the direction from where he had come.

"Maybe he is right," said Mac. He placed his hand on Lise's. "It does matter what happens to you. So maybe we should wait for Eve." Lise's eyes blazed with a brewing anger. "We don't know if Eve will ever be able to do this type of interface again," she spat as she pulled her hand away. "And even if she can, there is no reason why she should have to take all the risks. She already did when she interfaced with the satellite. It's time we start sharing these risks. And I can do it."

Lin waited until she was sure Lise and Mac didn't have anything else to say for the moment. "You will need information, right? So you know what to look for?"

"The more the better," said Lise.

"Then we should tell you what we have learned since we arrived." Lin glanced over at Mac who was still struggling with Lise's outburst.

"That would be helpful."

Lin pressed a couple of buttons on the tabletop and a display rose from underneath its surface. Once the display came to life, Lin manipulated some of the symbols depicted on its face before turning the display towards Lise.

"When we arrived in this cavern, we discovered what we later determined were two sets of remains of some sort of creature," started Lin. "They were surprisingly well-preserved. I was able to retrieve a complete set of their protein schematics and recreate virtual copies of both organisms. They were from the same species, one 'male' and one 'female'. Here is what they look like."

The display swirled briefly before two diagrams appeared in front of Lise. The figures were humanoid, each having two arms and two legs. They were tall and thin and had long segmented fingers. A ring-shaped organ encircled their heads giving the appearance of a thin crown. They had large eyes and drooping mouths that barely had any chin at all.

"What are they?" asked Lise.

"No idea," said Lin. "Our analysis was significantly impaired initially because we didn't have access to the equivalent of an Alpha panel. But Mac was able to rework one of the panels we found here so it could serve our purposes. We also didn't have access to a decent Data Agent sim. But despite that, Mac and I were able to determine quite a bit about their biology. They have a rather unusual reproductive strategy."

"Are there any on this planet?"

"Not that our scans have detected," said Lin. "But that doesn't necessarily rule it out. They reproduce sexually, but their bodies lack the various sorts of enzymes that we usually see in organisms that reproduce like that. These enzymes enable the male protein structures to unfurl so they can bind with the female's protein structures. Since these enzymes are missing, they have to manage this in another way."

"Which is?"

"They require an extraordinary amount of energy to bring the male and the female contributions together. Typically enzymes would allow this to happen at lower energy levels."

"But why were they here then?" asked Lise. "This planet doesn't have much energy of that type."

"It used to," said Mac.

He now had Lise's full attention. "What do you mean?"

"Lin, this is consistent with the data you had me review," said Mac. "This organism apparently was well-adapted to the environment this planet used to have."

"But not always," said Lin with a burst of enthusiasm. "Remember? The data said that before Haven suddenly heated up, it was much cooler. You said that some intelligence had to be responsible for the sudden increased temperatures."

Mac nodded. "Of course. It has only been the passage of time, a lot of time, that has resulted in the planet cooling off again. These aliens must have found Haven during the period when it was much hotter and decided to stay here."

Lin was rolling the possibilities around inside of her head. "That would make sense. But we still don't know who heated the planet up in the first place."

"This species found an almost perfect planet for reproduction," said Lin. "The temperature was hot enough to allow them to exchange their protein templates, but....."

"'But what?'" asked Lise.

"Everything else about the environment was right as well. We don't know what they eat, but there must have been food for them here. And if the atmospheric gasses weren't right for their metabolism, the temperatures and food supplies wouldn't have mattered. The odds of an alien species just happening along and finding such an ideal planet are extraordinary. But I haven't found any fossil evidence suggesting that this species evolved here on Haven. It's miraculous. But that also explains the number of offspring."

"What do you mean?" asked Lise.

"This species reproduces sexually, but the product of the reproduction looks a lot like a seed." Once again she manipulated the display. Lise could see a large circular structure with a number of tentacle-like appendages extended from it. "And there are a lot of them. Likely thousands."

"So where are they?" asked Mac.

"Maybe that was what Eve was trying to warn us about," said Lise. "Maybe they are still here on the planet and we just can't find them for some reason."

"Have you finished your study of the organism?" asked Mac.

Lin shook her head. "Not quite. These panels and displays aren't exactly like ours. We also didn't have access to a Data Agent, or a suitable sim personality, to help us analyze our data. There will be additional information about them in the panels that we haven't been able to retrieve yet. I was hoping you might do that while you are doing your interface, Lise."

"Lin, this technology, at least from what I can tell, is very similar to the technology we have on our Journeycraft. How did they get it?"

"It is very similar," confirmed Mac. The panels, the displays, even the Journeycraft that are littering the area. All of this technology had to come from the same place."

"So Journeyers have been here before," said Lise. "But what happened to them? Why didn't they come back?"

Mac walked over to a nearby bank of displays. "All of these were here when we arrived. None of them were powered up because they drew their power from an outside source. We managed to get a few of them functioning. There appears to be one that is central to all of the others."

"Mac," said Lise, "did you all do a surface scan before you landed?"

"Jack did, I think, but he didn't share any of the information. Why?"

"Sim-Min said he found twenty-nine damaged structures on the surface that were consistent with beacon satellites that had fallen back to the planet."

Mac's eyes grew wide. "The beacon satellites are typically the power source for these panels. So when they stopped working, the panels stopped working too. But here's something really odd. Guess how many panels we found in here?"

Lise swallowed. "Twenty-nine?"

Mac nodded.

"Jack said there were messages going back and forth from Uma to Haven," said Lin. "From our beacon satellites to these." The final realization struck all of them simultaneously. "Uma was communicating with the aliens."

"We need to access those communications then," said Lise. "The sooner the better. I'll do the interface and get as much data as I can."

"That might be overly broad," said Lin. "You don't want to end up like Eve. Or worse. That wouldn't help anyone."

"Maybe I can limit it to simply communications then instead of information generally," said Lise. "Jack said that messages were being sent to Uma from Haven. If I can find those messages, maybe we can figure out the connection between the two places."

"That makes sense," said Lin. "I'd like to review Eve's status," she said, rising from the table. "And I want to examine you before your interface. There is no sense in hurrying."

"What if they are out there?" said Mac.

"So what if they are?" replied Lise. "If they had plans for us then I'm thinking they would have shown up already."

"Are you ready?"

Lin stood behind Lise who was seated in front of one of the larger panels. Lise wiped at a bead of sweat as it crept down her cheek.

"I don't know," said Lise. "I hope so."

"Are you sure.....?" began Mac.

Lise turned towards him and scowled before facing the panel again. "Yes. I'm sure. It has to be done. And I can do it." The panel seemed to emanate a deep cold, as if she were staring into space itself. Lin completed a quick assessment of her health.

"Neurologically, you are as fit as you can be," offered Lin. "If that makes you feel any better."

Lise attempted a smile. "It doesn't," she admitted. "But I'm still a long way away from being a fern, right?"

Mac muttered under his breath.

"I guess Jack isn't coming," said Lin. "Not surprising. No one can do anything that he can't do. There's no point."

"Lise can," said Mac as he placed a comforting hand on Lise's shoulder. "Just watch."

Lise closed her eyes and took a deep breath. "Messages exchanged between Uma and Haven," she said to herself. "Most of the data transferred to Uma would likely just be routine data downloads from Journeycraft, so I can try to screen them out. I'll try to focus on just exchanges originating from the planets themselves."

"Yes, that's an excellent approach," said Lin. "Do you have any idea about how long this process takes?"

Lise shook her head without looking away from the panel. "No, not really. It's different for every Data Agent. And this is my first try."

"We will be here waiting for you when you are finished," said Lise. She glanced at Mac who hesitantly removed his hand from Lise's shoulder. "Ready when you are."

Lise took another deep breath and placed her hand on the display. At first there was no response, but soon her hand was outlined with brilliant white light. Lise's body tensed and her face grew still with concentration. Lin stood by trying to assess her vital signs.

"She's doing great," Lin told Mac. Mac couldn't draw his eyes away. Lise was frozen in position, her eyes closed and her expression drawn taut with tension. A fern? Mac hated Jack for placing that thought into his head.

Lise remained completely static for an indeterminate amount of time. Lin and Mac weren't sure what to do. They were afraid that any sudden movements or noises might break Lise's connection with the panel. The two of them stood quietly attempting to remain as still as possible.

Finally, Lise drew a long, deep breath but her eyes remained closed. The tense expression on her face transitioned to fatigue. Sweat began to drip down her forehead and onto her face. Lin cautiously slipped an examination ring onto her arm and placed her hand on the back of Lise's neck.

Lin nodded. "She's fine," she whispered. "Her neural connections are firing very quickly but her body is holding up very well. Her pulse and blood oxidation are excellent. She's really doing it."

Mac smiled as his anxieties eased. Lise opened her eyes and removed her hand from the display. She turned in her chair and stood up. The three of them hugged, the strength of their embrace drawing them close enough to feel each other's relieved laughter.

"How do you feel?" asked Lin.

"Odd," said Lise finally. "The entire experience. Odd. It was like watching a series of events that were happening at incredible speeds. But my mind, somehow it was able to change the speed of the events I was seeing so I could focus on what was important."

"Did you see anything important?" asked Mac.

Lise sat back down and bade the others to do the same.

"As we suspected, the vast majority of the data going towards Uma was simply updates from Journeycraft. In fact, I didn't detect any information being transferred directly from Haven to Uma until I was nearly finished. The data from Uma to Haven is rather routine as well. All of the historical information placed into the panels by the Chroniclers was sent here. It's as if someone or something on Haven was keeping up to date on what was happening on Uma."

"You said you found something from Haven to Uma right before you finished," said Mac. "What was it?"

"It was one of the very oldest messages," said Lise. "It was a set of instructions stored in every Chronicler Hall panel on Uma." Her expression went blank and she stared forward.

"Lise?" asked Mac. "Lise?!?"

"They were watching us.....," she said in a monotone. "They want Uma for themselves. They've sent........their offspring there already........"

Lin pushed Mac aside and returned the examination ring to her forearm. Once the membrane was in place over her hand, she placed it on Lise's forehead. "Her neurons are firing at a heightened rate again," she said quietly.

"But she isn't connected to the panel anymore!" protested Mac.

"No, but there must have been more information that her brain wasn't finished organizing. It's doing that......."

"Agggghhhhhhhhh!!!!!!!!!" Lise screamed as she covered her face with her hands. "It isn't right! I've made a mistake!"

"Lise," said Mac, attempting to remain calm, "what is it? What did you see?"

Lise drew a deep breath. Slowly exhaled, and then closed her eyes. "That pair of aliens, they were a brood pair. At a certain point in their physical development, they traveled through space looking for suitable breeding grounds. Their ships….they looked a lot like our Journeycraft. They were looking for planets inhabited by intelligent beings that could be manipulated."

"That explains why they had so many panels here," said Mac. "They must have been monitoring planets from a large number of systems in every direction. But what do you mean by 'manipulated'??" asked Mac.

"To drastically increase the degree of radiant energy in the planet's environment."

"Yes!" confirmed Lin. "To reproduce. They need a high degree of radiant energy to reproduce."

Lise lowered her eyes. "Then I did do the interface correctly. I'm not mistaken."

"What else, Lise?" demanded Mac. "Mistaken about what?"

Lise drew another deep breath and stared vacantly at the panel she had accessed previously. "These creatures. They are exceptionally long-lived. The brood pair traveled to various host planets and staged them so they'd be ready for when the pair's offspring arrived. Each staging was different depending on the type of society they initially discovered." She drew her arms around her chest and shuddered.

"Lise?" said Mac quietly. "Are you……"

"Mac," she said quietly, "I saw Uma. They sent offspring to Uma."

"But…..how?" asked Lin.

"The tether beams. We thought they were intended to allow for the transfer of information from the Journeycraft to Uma and for navigation. Although they did that, they also provided a pathway for their offspring."

Lin sat back, a wave of defeat washing over her features. "Stars. The signals are nothing but ordered energy. It would have been like a food trail for them to follow. And those satellites, those signals, were all maintained by the Journeyers."

"But they couldn't have known!" said Mac.

"No, of course not," said Lise. "They were attempting to find a new home for us. They were trying to find this place."

"But how can they propel themselves through space?" asked Lin. "It would take them forever to travel that far."

"Our Journeycraft," said Mac. "They followed the tether beams as well. And the tether beams aren't that large."

Lin allowed Mac's comments to roil around inside of her head for a moment. "So, the offspring got a ride in the Journeycraft?"

"Or on them," said Mac soberly.

Lin turned back to Lise. "What about the Directors? Did you see anything about them? This planet HAS to be Haven. I thought for sure they would be here. Or at least some remnant of them."

"No. I didn't find any information about the Directors," said Lise. Fatigue weighed heavily on her speech.

Lin gasped. "My stars, for all we know, the offspring are already there! On Uma! They are a new brood pair. We need to finish that analysis of their protein templates to figure out what they are going to do and how we might stop them."

"There's one other thing," said Lise. "The instructions I mentioned. They were distributed to every Chronicler Hall on Uma. But the instructions were designed to send a signal back here once they were accessed on Uma. Those instructions were sent to Uma tens of thousands of ellipses ago, but they were only accessed very recently. Once on Atla and once at the Land Bridge community. And those instances occurred within rotations of one another."

"Do you know what they say?"

Lise shook her head. "They had exceptionally high information densities. Each time I tried to access them I was nearly overwhelmed and had to halt my attempt."

"They must have had tether signals going in a number of different directions, towards all of their target planets. That could explain why so many of Uma's Journeycraft didn't return."

"Explain that, Mac," said Lin.

"The tether beams could have confused the nav functions of the Journeycraft, particularly while the crew was in repose. If a Journeycraft detected a second tether beam, it wouldn't have had a contingency response. That possibility was never contemplated by anyone on Uma. So those ships may have just changed course to follow the new beam. Journeycraft are directed to bring their crews out of repose sleep once the ship reaches a certain set of coordinates."

"And since the ship never reached those coordinates, the crew never woke up." Lin was nearly broken by the next step in Mac's logic. Mac nodded quietly.

The three sat in silence pondering the ramifications of this new information.

"We need to go back," said Lin finally. "To Uma. Even if the brood pair is there, we might still be able to stop them."

"How long before your protein schematic analysis is ready?" asked Lise.

Lin turned towards Mac.

"I'll go back and start working on the algorithms again," said Mac with a renewed burst of energy. "I'll do whatever I need to do to accelerate that process."

"And Jack?" asked Lise.

"What about Jack?" replied Lin. "Anything we can get done will have to be done without him."

"If we could revive Eve, we could finish our analysis a lot faster," said Mac.

"You still don't trust me to order the data for you?" said Lise.

"It's not that. But if we had Command Authority aboard the *Starshine*, we could do the analysis in the Life Lab there."

Lise jumped up from the floor. "Well, I don't have Command Authority, but I know someone who does. Get your data together. We'll have that analysis completed in no time."

CHAPTER 13

The clan did not know what to do with the dead. It was customary to place a recently deceased member's corpse on a pile of dry sticks and set it on fire. But now there were so many. They littered the area outside the cave opening such that it was nearly impossible to walk there without stepping on a corpse. So the clan members huddled inside the cave instead anxiously waiting for Dom to speak.

He let their anxieties ferment. This was an important moment and he wanted to be sure to capitalize on it perfectly. He rose and waited for the others to assume the role of his audience.

"We have won a great victory." He spoke in a near whisper. The Umae remained silent, leaning towards him to better hear his quiet words. "But it is only one victory. Only one of many we must achieve if we are to survive."

Joba was off to one side with a group of Hek acting as an interpreter. As he began to translate Dom's words, Dom raised his hand and bid him to be silent. "Not now, Joba," he said in a louder voice. "You can explain my words to them later. For now, tell them that they must go outside and remove the bodies. If they remain there in the sun for too long, they will draw predators. It isn't safe. They must remove all of them and take them far away."

Joba took a moment to make sure he understood Dom's request. "You want the Hek to go outside and remove the bodies?" he asked finally. "Isn't that a task we should all share?"

Dom's eyes narrowed. "The Sky Fire has risen. The Hek are better suited to endure its harsh flame. Ask them if they are stronger in the fire than the Umae. The Hek will tell you."

"But they are better suited for many things," protested Joba. "They have taught us much about life away from the pyramids."

"It is their skin," said Dom. Now he was speaking in the tongue employed by the Umae. "What happens when an Umae spends too much time in the fire?"

One of the Umae men spoke up. "It turns red, like meat. It hurts." The other Umae nodded in agreement.

"And what of the Hek?" Dom paused, allowing the Umae to draw their own conclusion.

"Their skin turns brown," said the man. "They don't burn as much." Again, the Umae joined in agreement.

"He wants the Hek to remove the bodies," said Joba to the Hek. "And take them far away. He fears they will attract predators."

The Hek glanced at one another before rising to their feet, almost in unison. "The child speaks truth," said another. "Come," he said, motioning to the other Hek in the chamber. "If we have more, we will return sooner." The group surrounding Joba, as well as the other Hek inside the cave, all rose and headed outside to begin their macabre task.

Dom's smile was nearly imperceptible.

"What about the rest of us?" asked an Umae woman. "We can't just sit in here."

"No, we can't," answered Dom in Umae. "I have important words for the rest of you." Now the volume of his voice was rising. "Were there any of you who did not see the battle earlier? Approach if you did not." Perhaps one out of five slowly made their way closer to Dom, each glancing around at those who remained where they were. Once they finished their approaches, Dom turned towards Joba. "Joba, did you see how Nagham and Tarak died?"

"Tarak was stuck with a spear in his neck," he said glumly.

"Who threw the spear?"

Joba thought for a moment. "I don't know. There was so much happening."

"And Nagham?"

"Nagham was seized by a large man. He.....crushed her head." The young man was barely able to form the words describing her death.

"A large HEK man," said Dom. "He took her by the hair and beat her head on a rock until she died." He had no problems describing the incident. "And just prior to that, what had she done?"

Joba thought back. "She had thrown a spear at a member of the other clan."

"Also a Hek," noted Dom. "And the individual who killed Tarak was a Hek as well. I saw it."

The Umae in the cave muttered to themselves, unsure as to what Dom was trying to tell them. "But there were Hek and Umae in that clan," said Joba. "Just like in ours."

The other clan members were nodding their heads in agreement.

Dom waited for them to become calm once more. Again, he spoke in a quiet voice. "Let me tell you what Nagham told me just before she attacked the Hek in the other clan. She told me that this clan had led her astray. It had forced her to tell you lies about where she had come from. And she had lied because she wanted to rule you as a 'gift'."

Protests rolled through the group. "But she was an elder!" said one man. "She provided us with much guidance and we prospered."

"'Prospered'", spat Dom. "Tell me, did you live with the pyramids before the Great Rain?"

The man nodded. "Yes, until I was a young man."

"And did you raise your stock, your cows and sheep and chickens?"

"No. I was a woodworker, others did those things."

"And did they confine those animals in pens and fields, with fences and cages, the entire time making them fatter until it was time for their deaths?"

"Yes, that was what they did."

Dom drew a deep breath, reconnecting with the audience. "And did you have to bear spears, or nets, or slings and go into the wild to kill these animals?"

The man seemed confused. "Why....no. We already had them in the community. Why would we hunt them? And besides, we did not kill them. It was forbidden. We only used their bodies after they died."

Dom hesitated ever so slightly. "So why do you hunt them now?" he asked.

"The Hek have shown us," said Joba. "They have taught us to hunt and to fish."

"I'm sure they have," said Dom. "And they have taught you to gather fruits, and nuts, and roots?"

"Of course."

Dom looked back at the woodworker. "Did you do that in your community?"

"No. We grew those things ourselves. We didn't need to look for them."

"And they taught you to kill," noted Dom. "Now you hunt and you kill, do you not?"

After a moment, the man nodded.

"The Hek have made you their servants," said Dom. "And you don't even realize it. You follow the Hek way, which is hard. Instead of raising your own animals, you hunt them as the Hek do. Instead of growing your own vegetables and fruits, you gather them as the Hek do. And they taught you to kill without giving you any guidance on when to use such a dark power. Tell me, how many of you have taken a Hek as a mate?"

A small portion of the crowd moved forward, almost all of them women. "What are you saying?" asked Joba. It was a question everyone in the chamber had wanted to ask.

"And how many of you have borne young with the Hek?" No one in the group responded. "Anyone?" The Umae in the chamber passed nervous glances.

Dom released a long, dramatic sigh. "Don't you see? They have been the enemy all along. For ellipses beyond counting they watched and waited. When the Great Rain fell they had their chance. Given the opportunity, they have enslaved you but you can't even see it." He released another sigh, this one followed by a healthy pause. "And this is our fault, the fault of the Journeyers. We made everything too easy for you. We showed you how to grow food. We provided you with shelter. And you gladly accepted all of those gifts. But they made you soft and trusting."

Joba was the first to find his voice. "But they have helped us," he protested. "They helped protect us. We knew nothing of living away from our homes. Without them….."

"What?" snapped Dom. "You would have died? Hardly. Have you so little faith? We saw your struggles. We stood ready to assist you as we always have. If you had not joined them, you would not have perished. We would not have permitted it. We Journeyers were deeply aggrieved by your decision. I had to plead with my fellows to allow me to come to you as I have to show you the mistake you have made. The Hek force you to use their ways. Perhaps you have forgotten your own? And they take you as partners despite the fact that you cannot bear young together." He sneered at a group of nearby women. "You serve them in every way possible and they take pleasure in your servitude."

An Umae man stepped forward. "But they can have young with us," he said firmly. "The Others. They are born of one Umae parent and one Hek parent. We have some among us in our clan."

Dom repressed the deep satisfaction that was rising in his breast. It was so easy to lead these people to the conclusions he wanted them to reach. "Tell me," he said in a quiet voice, "how long has it been since such a child was born?"

The man frowned. "It has been many ellipses," he admitted. "None since the Great Rain."

"They withhold this favor," said Dom. "As only the Hek can. It is a part of their plan."

"What plan?" asked the man nervously. "We know of no such plan!"

"No. You don't. That's why I came. It is a plan for your complete destruction. The end of the Umae."

He paused and allowed his words to ripple through the crowd. Individuals whispered frantically to one another, trying to understand.

"Once you forget your ways, they will cast you aside. You are physically weak. You are not suited for the life of a hunter. And they are preventing you from increasing your numbers."

"But their numbers aren't increasing either! There are no young Hek!" The rest of the crowd grumbled in agreement.

"Oh no?" asked Dom. He tainted his question with a dark lacquer of pregnant revelation and hesitated. "They are like leaves on a tree. I have seen them. Great droves of Hek biding their time. What do you suppose they are doing while they walk beneath the Sky Fire as you sleep? They are meeting with other Hek. They are making plans. After what I have done for you, why would you doubt my words?"

The man turned around and studied the faces of his fellow Umae. "This is very distressing," he said. "But we have no reason not to trust you. What would you have us do?"

The other Umae were quiet now, eager to hear Dom's words.

Once again, Dom struggled to quell his enthusiasm. "Listen carefully. I will tell you what must be done."

The Hek spent the rest of the day moving the bodies of the dead and burning them. Although they had determined that their fires were distant enough that the smoke would not reach the cave, they were wrong. After the wind shifted, the cave was filled with the sweet stench of burning flesh. The Umae within quickly gathered their belongings and fled in the direction of the wind until the smell dissipated. They huddled beneath the canopies of tall, leafy trees in an effort to shield themselves and their young from the burning gaze of the Sky Fire. As the light began to wane, the Hek found them and joined them next to the trees. They reeked strongly of the dark odor given off by the dead as they burned.

Dom approached them with a patch of cloth over his face. "You must move away from here," he directed. "Your odor is sickening. And you will draw predators to us. Go now! Find a river or a stream and clean yourselves. Return once the Sky Fire casts its first light in the morningside sky."

The largest of the Hek who had cleared the corpses frowned. His name was Palu and he was exceptionally well-built even for a Hek. A half-head taller than the others, his chest was as thick as a tree trunk and his shoulders were woven thongs of muscle. "It is safer if we stay together," he protested. "We have no shelter."

"You should have thought of that before you decided how you were going to burn the bodies," said Dom. "You lit the fires too close to the cave."

The giant Hek's brow furrowed. "But you told us to burn the bodies and come back," he said. "We did that."

"You did it wrong!" snapped Dom. "And if wild animals follow your smell here, you put all of us at risk. If you are away from us, we will be safe. You would put us all in danger with your own mistake?"

Palu considered Dom's logic. "We have always stayed together. The sick and the slow are bad for the rest, but we keep them anyway."

"But they didn't choose to be sick and slow, did they?"

Palu lowered his eyes. "I did not choose the place for the fires," he admitted. "I wanted to move farther away. Some of the others could not carry a body that far."

"It doesn't matter whose idea it was. You were all there and you all did it together. You all...." Dom abruptly stopped speaking. He then switched to the Umae tongue. "You dim-witted buffoon, you have given me an idea."

Palu searched for a nearby interpreter but none was about.

"Who picked the place for the fires?" asked Dom, once more employing the Hek language.

Palu pointed. "It was Racha. She could not carry the bodies far enough. She asked us to move closer." Dom followed Palu's gesture and saw a young Hek female sitting with a group of 4 other Hek.

"Tell her to come here," said Dom. "I need to talk with her. Alone."

Palu stared hard at Dom for a moment before moving off towards Racha's group. Dom could see Palu speaking with Racha before she rose and walked over to where Dom awaited her. She was plainly confused.

"You wish to speak?" she asked.

This was a meaningless moment for Dom, but one that he knew would come ever since he had accessed the Alpha Panel on Atla. He would simply get it over with before moving on to the next step in his plan.

"Do you understand Umae?" he asked. Racha stared at him blankly. "I thought not." He switched once more to Hek. "You are Racha?"

"Yes. I'm called Racha."

"Do you know who I am?"

Racha blinked a couple of times. "You are a magic person. A Gift maybe?"

Dom barely managed to suppress a laugh. "Is that how they told it to you?"

Racha nodded. "Some Umae can speak Hek too."

"Indeed. But let me ask you this." His voice became low and quiet. "Do you wish to serve me?"

Racha turned her head to one side. "Serve you?"

"Yes," he continued with the same tone. He redoubled his efforts to force his voice into her simple mind. "Do exactly as I say. Always. Without waiting for anything else."

"I do not understand," said the woman. "No one serves one person. We all serve one another."

There was no anger or disappointment. Dom had simply confirmed what he had known for a long time.

"You chose the place for burning the bodies?" He was now using his normal tone.

"Yes. My arms grew heavy. I couldn't walk that far."

"And now we have had to leave our cave."

Racha was unaffected. "We have left many caves," she noted with a shrug. "We can find a new cave."

"Yes, but it is dangerous. Animals can find us here. Find us and kill us."

"We can fight the animals. Just like always."

Dom rose and called to Palu. "Palu, bring the Hek who helped you burn. We have a problem." Palu nodded and began gathering other Hek to him.

"What is the problem?" asked Racha. "Are we going to look for a new cave now?"

Dom ignored her. He stood silent while the other Hek approached. There were at least 30 of them with Palu at their head.

"Your smell places us in danger," he said slowly. He wanted to be sure they understood. "We had to leave our cave because of the smoke from the dead. The fire was too close to the cave."

He waited to gauge their reaction to his message. It was indiscernible.

"Racha chose the place for burning," he continued. The Hek looked at Racha, many of them nodding in agreement. "Her choice has placed the rest of us in danger."

The Hek grumbled amongst themselves. "There is no danger," said Palu. "Our group is strong."

Dom shook his head. "No. No. Why should all of us be in danger because of a decision made by only one person?"

"What do you say?" asked Palu. Once more, Dom heard the exact question he was hoping for.

"She must go away until the Sky Fire returns. When she is clean, she may return. The rest of you can remain. We will need you to protect us."

"No!" cried Racha. "Anyone who is alone while the Sky Fire sleeps is in danger!"

"And it's not safe for us to let you stay here and attract animals," noted Dom. "Next time maybe you will try harder."

Racha looked about anxiously at the other Hek. "What if we go with her?" asked Palu. "Then she will be safe."

"But what of us?" asked Dom. "What of the Umae?"

"You will be safe without us," said Palu. "And Racha will be safe."

Dom pretended to consider Palu's decision. "Joba! I need Joba!"

The young Umae eventually appeared from within the group resting under the trees. "I'm here!" The Hek wondered why his presence was requested. Joba was equally as perplexed by the large group of Hek talking with Dom.

"I want you to listen to what Palu and his group are going to do," said Dom. "Palu?"

Palu took a step towards Joba. He addressed the young man in Hek. "Racha must go away until the Sky Fire returns. The Hek will go with her."

"And they are leaving the rest of us here, by ourselves," added Dom in Umae.

"But......" Joba turned back towards Dom. "I don't understand."

"You will," said Dom. "Go and tell the others to prepare to spend the night here. We will need to light fires and have people stand watches. We have no cave tonight."

Joba nodded his head before moving away. Once he was gone, Dom addressed Palu once more in Hek. "Now go. Make her safe. Return only when the Sky Fire returns. Only then will we be safe."

Once again, the Hek grumbled but Palu bade them to be silent. Gathering up their weapons, the Hek strode off into the fading light and disappeared into the woods.

Dom smiled as they left and began making advanced plans for battle.

Chapter 14

Lise, Lin, and Mac exited the pyramid and headed for the *Starshine*. They stopped at the *Aurora* on the way to see if Jack had done anything to provide them with access to the ship. Its ramp was still up and neither Lin nor Mac could do anything to make it drop. Lise made another effort at banging a rock against one of the ship's landing piers but that drew no reaction save the curious glances offered by her companions.

"Why won't he let us in?" asked Mac. "He's never done this before. I wonder what he is doing in there?"

Lise shrugged as she took Mac's hand. "I don't think logic will give us any insights into his behavior," she said.

"You are more right than you realize," said Lin. "Once we get to the *Starshine*, I'll explain why."

With one last glance up at the *Aurora*, the threesome quietly covered the remaining ground to the *Starshine*. As they approached, the egress ramp lowered permitting them to enter.

"Did you program that?" asked Mac.

"No," said Lise. "My temporary Command Agent did that."

"Temporary?"

"Just wait. You've already met him."

As they entered the ship they could hear various panels at work. A rapidly blinking light on the Alpha panel caught Lin's attention. Her eyes widened as she sprinted away towards a side passage. "The Life Lab! Come on!"

Lise and Mac dashed after her, turning a corner just as she was entering the Lab a short distance away. As they entered, they saw Lin standing next to one of the beds. She was preparing an examination circlet as Eve, sitting up in the bed, watched closely.

"Eve!" said Lise. "You are all right!"

Eve looked pale and haggard, apparent even though her skin was typically almost entirely without pigment. She tried to smile as Lise began her examination.

Lise paused and periodically moved her hands to different points of Eve's body, allowing the membranes on her hands to transfer information about Eve's physical status directly into her mind.

"She's going to be fine," she said finally. "What I would expect based on

Lise's description of what happened. That process must have been extremely taxing." She removed her hands from Eve and stripped the membranes away.

"I'm fine," said Eve weakly. "And how are you?" she asked, looking towards Mac and Lise.

"Don't talk," said Lin. "Rest. You need rest."

"We are fine," said Lise. "We are all fine. And we are glad that you are as well. It has been a long time."

Eve noticed Frell, suspended in his repose tube. "Is Frell.........?"

"He's fine," said Lin. "Really. Sol suggested a plan for Frell so we all decided to return him to repose."

"Frell agreed?" asked Eve.

Lin nodded. "He did. He understood. And after you have rested a bit, we will fill you in."

Eve put her head back on her pillow and closed her eyes. It wasn't apparent to the others if she was asleep or merely resting.

"We will leave her for now," said Lin. "I can monitor her through the panel displays. If she needs anything, I'll know." Lin took Mac and Lise by their elbows and guided them out of the room. Once the door slid shut, she stopped. "Her nervous system was severely compromised by whatever it was that she did. She's going to be fine eventually, but I doubt she will be able to do another interface like that one. She may not be able to do any interfaces at all."

"Are we going to be able to retrieve any of the data she accessed?" asked Mac.

"Mac! She's ill," scolded Lise. "What sort of a question is that?"

"It's an important one," said Lin. "She accessed an enormous amount of data. A lot of it may be similar to that which you accessed inside the pyramid, but the only way to know that is to find a way to retrieve it. We may not be able to do that for a long time, and maybe never at all."

"I could access it," said Lise. "I did the panel in the pyramid. I can do that beacon satellite as well."

"No!" said Mac. "You can't. I won't let you."

"It's not your decision......."

"Technically it's my decision," offered a disembodied voice from overhead. It was sim-Min.

"That temporary Command Agent I mentioned," said Lise. "He.....it..... has Command Authority."

Mac looked up at the ceiling. "But that's impossible. It's not even alive. The ship's protocols wouldn't allow it."

"How do you know that?" asked Lise. "Have you reviewed them?"

"No, but no one entering directives into a Journeycraft's primary system core would allow for such a thing."

"Not even the person who created those directives?" asked Lise.

Mac absently rubbed his chin. "Min wrote those directives, didn't he?"

Lise nodded. "Yes. He told me that he, or the real Min, designed the ship's core function directives to make sure that someone, or something, always had Command Authority. After Eve finished her interface with the beacon signal, she collapsed. Sim-Min determined that since Eve wasn't able to exercise her Command Authority over the *Starshine* that he, I mean it, should assume that authority."

"It was the appropriate decision given your lack of experience, Lise," said sim-Min.

"I don't disagree. But I'm wondering what Jack is going to think about this."

"I don't think I follow," said the voice. "Jack has Command Authority over the *Aurora*. It would be impractical for him to have Command Authority over both ships since he can't be aboard both simultaneously."

"I don't think he will see it like that," said Mac.

"Let me finish telling you what we began talking about earlier," said Lin. "Jack has DSM, Deep Space Megalomania, and it's severe. His judgment is extremely questionable. He shouldn't have Command Authority over anything."

"There won't be any way to change that," said Mac. "With all of his free time that we have been wondering about, I'm sure he has rewritten all the core function directives on the *Aurora*. In fact, he's changed just about everything on that ship. Its operational parameters are likely the most technically efficient of any Journeycraft ever. Its speed, maneuverability, processing capabilities, communication arrays.......everything."

"I'd like to review those specifications," said sim-Min. "But you are likely correct."

Lin sat down at a panel and began entering directives. "I'm going to enter the data we had about the alien protein schematics," she said. "Sim-Sol, are you still here?"

"I'm always here." It was a different disembodied voice, a virtual reproduction of the aged Life Agent's.

"Good. You can process this as fast as anyone, or anything. Review these schematics and tell us what you can about this organism. We already know about its reproductive cycles. What else is there?"

There was a pause as sim-Sol created the impression of consideration.

"This was the species responsible for the loss of the *Aurora*'s original crew," he said.

"How do you know that, Sol?" asked sim-Min.

"Jack and the actual Sol did an investigation of the *Aurora* and its shuttle crafts back on Uma. They found remnants of foreign protein schematics. All of that data was uploaded to the *Starshine* for storage. These protein schematics are essentially identical, varying only to a degree expected between different individuals of the same species."

"How can you conclude that that was what happened to the *Aurora*?" asked Lise.

"Several reasons. First, this species is drawn to radiant energy. The fusion engines and various scans emitted from a Journeycraft would be well-suited to draw its attention. Second, the Journeycraft was traveling within a tether beam, which is simply ordered energy. That would also attract this species. Third, this species seeks out intelligent lifeforms for the purpose of environmental management. The Umae would be suitable hosts."

"'Hosts'?" echoed Lin. "They are parasitic?"

"Not parasitic," said sim-Sol. "It would be more accurate to describe it as a dominant symbiosis. The conscience of the host is simply so subverted and repressed as to effectively absent."

"But how is that accomplished?" asked Lin.

"These creatures have internal organs with a highly unusual function," explained sim-Sol. "Both males and females. They also have a highly acute ability to quickly learn foreign methods of communication. I believe an individual of this species could communicate verbally with an Umae, using the Umae's language, in less than a rotation after first encountering the Umae. It would simply need to interact with the Umae and listen to it talk."

"So they are.....charming?" asked Lin, still confused.

"The organs I mentioned enable them to communicate with other species in such a manner as to effectively negate the will of the listening species. It is akin to hypnosis except more effective by several orders. These beings can, in effect, simply tell any member of the opposite gender of another species what to do and it will be done. It is essentially a neurological pheromone that is delivered by sound."

Standing on the steps with the sea far below, Lin had been in a hurry to reach the others at the bottom. She had a message to deliver. But Dom, as usual, was acting immature and wouldn't move along. He walked towards her and her brain swirled with fog. He wanted her to jump. She wanted to jump. As she fell, he seized her arm and pulled her back.....

"Lin? Lin? Are you all right?" Lise moved next to her friend and put an arm around her trembling shoulders. Lin buried her face in Lise's shoulder as her sobs shook her until she slid to the floor on her knees.

"I know," she gasped. "I.....know."

"You know what?" asked Lise in a soft voice. "Tell us."

"I know who the brood male is. It's Dom."

Lise placed a gentle hand on the back of Lin's head. "And the female?"

Lin slowly shook her head. "I don't know."

"Lin, how do you know it's Dom?"

Lin drew a deep breath and attempted to calm herself. "Because he did it to me. He told me to do something, to jump off a cliff, and I wanted to do it. It's him. I'm sure."

"The Directors, though," said Mac. "They aren't here. Where are they?"

"We may never know that," said Lise. "If they were here, they are gone. And we don't have any way to find them. We are on our own now."

Lise held Lin close while Mac stood by, mostly uncertain about what to do. Finally, sim-Min broke the silence. "You need to return to Uma," it said. "Immediately. The Umae race depends on it."

"What if we are too late?" asked Mac.

"You can't know until you get there," said sim-Min. "But if both members of the brood pair are there and have taken steps to transform Uma, you will need to gather as many Umae as possible and come back here."

"Here? Why here?"

"Because Haven, as it turns out, is the first planet suitable for Umae colonization on a large scale that any Journeyer has ever located. It is well within the parameters for climatic modification. And you have two Journeycraft."

"But we will need one to get to Uma, and one for the modifications," noted Mac.

"And I've seen Lin's analysis of Jack's condition," said sim-Sol. "He will certainly insist on using the ***Aurora*** as the ship to return to Uma."

"The ideal division of personnel would leave Lise here with us," said sim-Min. "We sims can do a lot of the work, but we can't do everything. There are reasons for deploying live Umae in Journeycrafts after all."

"No!" said Mac firmly. "If Lise is staying, then so am I." Lise glanced up at him with adoration as her eyes began to well.

"Then let me offer you a bit of advice," said sim-Sol. "Tell Jack that you think you should go on the *Aurora*. He'll disagree just to be disagreeable, and then you can stay."

"Jack wouldn't let you make any meaningful technical decisions anyway," said sim-Min. "So your utility would be much higher here. We would be working very closely."

Lin pulled herself to her feet. "We are going to need a way to sell this to Jack," she said. "Maybe we can make him think it was all his idea."

"And share this with him," said sim-Sol. "When these brood pairs send their offspring out into space, the offspring are housed in protective protein shells. For the most part, they aren't conscious as their nervous systems develop. But during that development they periodically gain and lose consciousness as various basic biological drives stir them. It is not unlike repose sleep."

Lin's face brightened. "And they have evolved to survive repeated awakenings!" she declared. "They are the key to our RTS research!

"Yes," said sim-Sol. "There are compounds generated by these creatures that greatly reduce the stress of these transitions. I find it highly likely that some of them could prove very useful in your research to cure Repose Termination Syndrome."

"That must have been what Sol, er, real Sol, was talking about," said Lin. A renewed vigor pulsed through her.

"What does the RTS research have to do with Jack?" asked Lise.

"Because if Min is still alive he is in repose on the moon," said Lin. "Don't you think Jack would just love to show his old teacher all the ways the *Aurora* surpasses the *Starshine*?"

"We can make him think it was all his idea," confirmed Mac. "And sim-Min can access the *Starshine*'s communications array so we can get him a message inside the *Aurora*." The group exchanged anxious glances. "Let's do it."

Jack emerged from his hidden chamber barely taking notice of the collection of Journeyers waiting for him. Sim-Sol had been correct. The information concerning the aliens and their plans for Uma had interested Jack enough for him to allow the others back on board the *Aurora*. He sat down in the Command Agent's chair and regarded them with an expression of extreme boredom.

None of the others were sure what to say. Jack had requested all of the data they had assembled and Mac had provided it. By now Jack had had enough time to thoroughly study it. He invited them all on board, and now……. Nothing.

"What are we going to do?" asked Lin finally. "We have a list of issues to address…….."

"Shhhhhh!" spat Jack. "Don't talk. Listen." Lin's face reddened but she did as he said. Not because she respected his command. He was apparently going to make a statement about what they had discovered. As much as she despised him, she couldn't deny his genius. She wanted to know his opinion.

"Fine."

"The Umae are in need of salvation," said Jack. He had waited momentarily before offering this pronouncement. He was feeding off of their interest. "Fortunately, I'm on their side. Of course, we will take the *Aurora*. The *Starshine* is laughably slow and I'm sure our people aren't interested in waiting for it. I already know how to deal with Dom."

"How?" asked Lise. "What are you going to do?"

Jack smirked. "It doesn't involve any of you. Don't worry about it."

"How can you say that?" asked Lin. "You are the Command Agent of the most powerful Journeycraft in the history of our people and have the opportunity to crew it with Umae who have undergone the most advanced training we have available. How can your decision to just do it all by yourself be the right choice?"

Jack folded his hands and placed his index fingers on his lips. "Humor me. What is your plan?"

Lin swallowed before sitting up straighter. "We should divide our personnel. Some of us go back to Uma on the *Aurora* to deal with Dom and his Brood Queen, if he has one, or to prepare as many Umae as possible for relocation back here if that isn't possible. The other group remains here with the *Starshine* so Haven's environment can be modified to optimize it for Umae habitation."

Jack snickered. "A fallback plan? So little faith. Lin, we won't need to modify Haven. We won't need to modify Haven because I'm going to defeat Dom and his Brood Queen, if necessary. The Umae will continue to live on Uma. There is no need for contingencies. There can only be one outcome."

"Oh? And then what?" said Lin. Her temper was rising. "You defeat Dom and his Brood Queen, and then what? Dom has already been there for hundreds of ellipses. Who knows what sort of changes he's effected? You can't plan for something that is completely unknown."

"Then I'll fix it," said Jack. "Whatever it is, I'll fix it. If he has destroyed the planet's ecological balance, I'll fix it. If he has tainted the Umae genetic foundation, I'll fix that too. This is not all that complicated."

"I can't wait to help," said Mac.

Jack turned towards his student and stared right through him. "You are, and will be, a waste of resources," he said coolly. "I'll handle all of the technical issues. There won't be anything for you to do, so there isn't any point in devoting any resources to you. I'm assuming that Eve will want to go so she can see Min again. Even though she may not be able to interface with any systems external to her, I'm confident she'd be our most useful option for data analysis. Besides me, I mean, but I'll be busy. So Lise, you are irrelevant as well. Maybe you and Mac can finally be together and tour the universe in that sad excuse for a Journeycraft. I really don't care what you do. You won't be on my ship."

Mac held his tongue. As much as Jack's comments angered him, sim-Sol had been correct once again. They had managed to let Jack conclude that all of this was his idea. Mac and Lise would finally be together. They just wouldn't be touring the universe.

"I'm going with you," announced Lin. "On the *Aurora*. You need a Life Agent."

"I do? Why is that?" Jack's tone dripped with condescension.

"The compounds generated by the aliens," said Lin. "Did you miss that part? I think there is a high likelihood they will lead to a cure for RTS."

Jack mulled over her comment for a long moment. "And why do I care about RTS?" he finally asked. "After I reclaim Uma, the Umae will never need to go back into space ever again. RTS isn't relevant."

"You plan on staying on Uma?" asked Lin. "Doing what?"

"Putting the Umae back together," said Jack plainly. "They won't have Protocols. They won't have Chroniclers. They won't have Journeyers. They won't have any ideas about how to live. I'm going to show them."

"What if they don't want you to show them?" asked Lin.

Jack shrugged. "Irrelevant. They will need me. I'll do what I need to do. I'll have all the technology in the universe. They won't have a choice. Fortunately," he added with a grin, "I'm benevolent."

Lin struggled to prevent her feelings of panic from making their way to her face. Jack wasn't paying any attention to her anyway. Her brief eye contact with Mac and Lise told her they were experiencing the same terror. She didn't need to formally examine Jack again. His DSM was raging.

"But Min is still alive," said Lin. "As is Sol. If I were able to do research on these compounds during the return trip to Uma, I might be able to save both of them." She paused for effect. "Wouldn't you like to see Min again?"

She had regained Jack's attention. His mind appeared to be cycling through all of the benefits of such a meeting. "You are that confident? I'm sure he would like to see the **Aurora**," he said finally. "After all, it is far improved from anything he ever put together."

"And I'll need Sol as an experimental subject," she added. Mac and Lise both started but remained silent, trusting enough in Lin that this was all a part of her plan.

"Fine. He'll be in repose the entire way anyway," said Jack. He looked over at Mac. "No resources necessary."

"So what now?" asked Mac.

"Now you all get off of my ship," said Jack. "Lin, you will need to make arrangements to transfer Eve and Sol from the **Starshine** to the **Aurora**. And as for you two," he said looking at Mac and Lise. "Hopefully Eve can transfer Command Authority of the **Starshine** to one of you. Because if she can't, it won't be of a lot of use to you."

"I'll talk to her about it," said Lin.

"One other matter. I want Frell on the **Aurora** as well," said Jack.

Lin froze. She desperately wanted that as well but Jack's request chilled her core. "Of course....." she began slowly. "But......."

"No 'buts'," said Jack. "I've seen the analysis you and Sol did, despite your efforts to hide it from me on *my* ship. I'll need a successor. One that I can tutor personally. If he is as smart as you seem to think he is, he is the only worthy option."

"Tutor him for what?" asked Lin.

"Sovereignty!" said Jack enthusiastically. "The Umae will flourish like never before and all because of me. But even I won't live forever. He will learn the ways of rulership, benevolent naturally, and he can only learn that from me. So now, the **Aurora** departs in one rotation. If you aren't here, I'm leaving without you. I'm going to go play hero and I really don't want to be late."

Lise, Mac and Lin gathered around Eve's bed as she regained consciousness. The Data Agent yawned and then frowned at her companions. "I'm guessing you are here for a reason?"

"How are you feeling?" asked Lin as she prepared to don an examination circlet.

Eve lifted a hand in protest. "I'm fine. Put that thing away."

"She's fine, Lin." Sim-Sol's calming voice sounded from overhead. "Given the number of tasks necessary for our return to Uma, I thought I'd examine her for you. I hope you don't take offense."

Lin frowned. "No, of course not."

"And based on sim-Sol's evaluation, Eve can resume her position as Command Agent of the **Starshine**." This time it was sim-Min. Eve looked up at the ceiling and fidgeted uncomfortably at the sound of his voice.

"'Resume'?," asked Eve. Given that sim-Min had no corporeal form, she looked to Lin for answers.

"You were incapacitated," continued sim-Min. Lin was grateful that he had continued his explanation. "So I assumed the position of Command Agent. But now that you have recovered, you are by far the best choice for the position. I have already re-established your Command Authority."

Eve was dumbstruck, her words frozen by a combination of surprise that a sim could even do what sim-Min had done and a moving appreciation for his kind words. Even though it wasn't really *him*.

Eve swung her legs over the edge of the bed and stretched.

"So? What's going on?"

"You are going back to Uma," said Lise. "You and Lin and Jack and Sol."

"And what about you and Mac? Where are you going?"

Mac took a half side-step closer to Lise. "We are staying here. We are going to modify Haven's climate in case you need to transplant the Umae here."

"Hang on here," said Eve. "Back up. What are you talking about? Why do you think we might have to transplant the Umae back here?"

"Let me explain," said Lin. "We have done a lot of research and analysis. Research and analysis on Haven, the data contained within Haven's panels, and on the creatures who used to live on this planet. The conclusions are disturbing."

"Eve has any of the data from the satellite surfaced in your mind?" asked Lise. "That was an enormous amount of information."

Eve drew a deep breath and laid back on her bed again. "Not much," she said somberly. "My mind must have suppressed it. That is a typical neurological response in a Data Agent when too much data is accessed at once. It will eventually come to me, but I may not be able to control the order or the rate at which it does so. More likely it will manifest as a stream of jumbled images and information that we will have to piece together." She looked down at the cover on her bed. "I'm sorry I didn't do better."

"Do you remember interfacing with me right after you interfaced with the satellite?" asked Lise. "You were terrified about Haven."

"No, I don't remember any of that," said Eve. "Did the results of all your analysis justify my concerns?"

"Haven's environment has already been modified, except in a manner that would have made it less suitable for the Umae," said Mac. "And that happened a long time ago."

"Modified? By whom?" asked Eve.

"By the creatures who lived here," said Lin. "We were able to recreate their protein templates and generate a virtual facsimile of the species. They are essentially planet-hopping progenitors. They go from planet to planet manipulating the indigenous species into modifying the planet into something suitable for these aliens to use as a breeding ground. They require a great deal of radiant energy in order to reproduce. Haven used to be very much like Uma, but then they changed it into a high energy brood world. Haven has slowly been reverting back towards its initial environment but it's a slow process. Lise and Mac are going to use the **Starshine** to accelerate the restoration."

"Why can't the Umae just stay on Uma?" asked Eve. The tension in her face indicated that she already suspected the reason.

"Because the aliens are already on Uma," said Lin. "Dom is the Brood King. We don't know if there is a Brood Queen there or not. If there is then we will likely have to move the Umae here."

"Dom? But he's just a boy!"

"He WAS a boy," said Lin. "At some point the alien Brood King assumed his body and suppressed his consciousness. I know. I encountered him when we went to the Land Bridge."

The weight of this information drove Eve back into her pillow.

"Can he be stopped?"

"Jack thinks so," said Mac. "He says that he has a plan."

"But he's not well," said Lin. "His DSM is worsening. But we don't really have any choice other than what he plans to do. The **Aurora** gives us our best chance to accomplish this in the shortest amount of time, and time is critical. And Jack has Command Authority over the **Aurora**. Even if we could demote him using the medical protocols, we still need him. He's mad but he's truly an exceptional genius."

"What do you need for me to do?" asked Eve. "I'm rather limited now and likely will be for some time."

Lise took her mentor's hand. "You are still an extremely skilled Data Agent and you have more experience than any of us."

Eve sat up abruptly. "What about Sol?!?

"Sol is in repose," said Lin. "He has RTS and it isn't safe to bring him out just yet. But our analysis of the aliens provided us with some insights that may well lead to a cure for that condition. Once we get to Uma, we might be able to cure him completely."

"Which leads us to the other part of our plan," said Mac. "Hopefully Min can be cured as well. He might be in repose at the Citadel. If Lin can find a cure, and we can bring him out we won't need Jack......."

Mac was stilled by the flood of tears rolling down Eve's face.

Lise gently squeezed her hand. "Eve?"

Eve drew another deep breath and tried to wipe the tears from her face.

"Min's dead," she said in a quiet, husky voice. "I was the last member of the *Starshine*'s crew to go into repose. His life sign signal terminated while I was alone. I'm sorry. I should have told you but I was concerned about morale."

A pall fell over the others. Lin absently ran her fingers through her light hair while Mac and Lise stood in stunned silence.

"Eve, I'm sorry," said Lin finally. "I'm sorry for your loss and I'm sorry we no longer have his guidance. But we still need you on the *Aurora*. Will you come with us?"

Eve stared straight ahead for a moment before nodding her head. "He'd want me to," she said finally. "And he deserves a suitable return ceremony. I can do that for him."

Lin leaned in and hugged her. "You can do something for all of us," she whispered as she kissed Eve on the forehead. "We need to get everyone transported to the *Aurora*. Mac, can you direct the mechs to get all the necessary equipment moved for us? I'll arrange to have Sol and Frell moved over."

"Absolutely," said Mac.

"And Eve......" Lin began.

"I'll walk," said Eve as she rose from her bed. She gripped Lin's arm for a moment to steady herself. She then stood up straight and lifted her chin. "At least I'll get to see him again. And we Journeyers aren't done with our job yet."

Eve directed the mechs from the *Starshine* to transfer her belongings and equipment to the *Aurora*. She insisted on walking herself despite Lin's pleas to the contrary. Although her steps were slow and measured, Eve made it to the other ship and even managed to scale the egress ramp without assistance. Jack was seated at the Alpha panel as Eve and the others entered.

"You and you," he snapped at Lise and Mac, "aren't you supposed to be away playing lovestruck nomads. Off my ship!"

Ignoring him, Eve turned towards the two young Journeyers.

"Mac, I have transferred Command Authority of the *Starshine* to you. She's a fantastic ship and holds many wonderful memories. You will find her more than able to execute the plans for climatic modification." She placed an arm around Lise's shoulders. "No offense to you, Lise. You have been a tremendous student and I know you will one day be a tremendous Command Agent as well. It's just that the C.A. of the *Starshine* has traditionally been a former Tech Agent. And besides, I was pretty sure you wouldn't mind."

Lise hugged her back. "No, of course I don't mind. You have made the right choice. Mac and I, and the *Starshine*, will have this planet ready in no time."

"And then what?" asked Lin. "You have so many options. Research, exploration.......what do you think you will do?"

Lise blushed slightly.

"Repose," said Mac firmly. "We've talked about it. We want to be around when the Umae arrive. Even if things on Uma aren't as grave as we suspect, someday the Umae will come here. There's no reason not to, and there are likely a great deal more secrets we can unravel about the aliens and their technology. Maybe we can learn what happened to the Directors."

"Well, you certainly aren't going to be able to do it," noted Jack. "Sleeping might be the most efficient way that you can contribute. But you may be out for quite a while. There won't be any reason for anyone to ever come back here. Everything on Uma will be better than it has ever been. I'll assure that."

Lise turned towards Jack but then decided to bite her lip. Mac was not so inclined.

"Jack you may be a genius, but you are by far the most miserable excuse for an Umae imaginable. Regardless of what you may or may not accomplish on Uma, you are just one person. You won't live forever. And, one way or another, the Umae will continue on. Our people are more important than any one individual, even you."

"Perhaps," chortled Jack, "but without me, there won't be any Umae. What they need is a figure to worship. To adore. I'm going to give them one. Now," he added with a more ominous tone, "get off my ship before I activate the security protocols."

Mac joined Lise and Lin in their embrace and then took Lise by the hand. "Grand journeys," he said somberly. "I happened with you."

"And we with you," replied Eve and Lin.

By the time Mac and Lise made it to the bottom of the egress ramp, both were in tears.

"Can we do this?" asked Lise as she looked around at the dusty ground and the nearly cloudless sky.

"Yes. Absolutely," said Mac as he gave her hand a squeeze. The egress ramp began to rise up towards the belly of the *Aurora*. "But we should move away. I wouldn't put it past Jack to try and incinerate us here on the spot."

The two sprinted away from the Journeycraft as the sound of its charging thrusters filled the air. As the thrusters engaged, the *Aurora* slowly began to lift off Haven's sandy surface before streaking into the light blue sky and disappearing. In the distance, the *Starshine* sat directly between them and the gigantic pyramid that was still so full of mysteries. Lise and Mac turned towards one another and embraced. The time had finally come for them to begin exploring mysteries all their own.

Chapter 15

Once the *Aurora* left Haven, Jack disappeared once again into his private chamber. Lin didn't mind and barely noticed. She devoted herself to structuring the experiments she wanted to execute using the chemical compounds generated by the aliens' simulations as the basis for a cure for RTS. Initially Eve wasn't able to help. Lin had essentially ordered her to remain on bed rest for fifteen rotations. By the twelfth, Eve was in much better spirits and was anxious for something to do. Lin refused to relent, and by rotation fifteen Eve was following Lin around the lab like a lost puppy.

Finally Lin allowed Eve to review some of the data analysis methodologies she wanted to employ. Eve quickly set to correcting them and to suggest entirely new strategies Lin hadn't considered. Jack had begrudgingly granted them access to sim-Jack. Jack had deleted all of the sim personalities from the *Aurora*'s panel except his own. "His" personality proved to be every bit as pleasant as the real Jack's.

"So you need for me to design a method of manufacture for the creation of a drug you haven't even developed yet, is that right?" it asked snarkily. Sim-Jack even had his sarcastic tone of voice.

"That's correct," said Lin calmly. "I'm sure you are familiar with contingencies? We have a broad array of generalities that we know will apply. What we need for you to do while we are in repose is to adjust the manufacturing methods to match the test results. I'd prefer to have Eve do it, or use her sim, but Jack won't let us. You are all that we have."

"Well, then you are in luck! I'm much more efficient than either of those options anyway and the odds of your project being successful will be much greater because of my involvement."

Lin looked over at Eve and rolled her eyes. "Is sim-Eve still available?" asked Eve. "You may well need access to a Data Agent profile while doing this work."

"Please. I won't need anything. And sim-Eve has been deleted. I'm the only simulation available and the only simulation you will need."

Eve sat down next to Lin. "The absence of a Data Agent's involvement in this entire process decreases the odds of success. I've run the numbers. The disadvantage is significant."

Lin sighed. "I was afraid of that. The sim seems to have tracked Jack's actual mental state. It's as crazy as he is. That overconfidence is starting to adversely affect our work. But I don't see any other options."

"I have one," said Eve. "I could just not go into repose."

Lin's eyes widened. "No. That's impossible. Besides we won't even reach Uma for another five hundred ellipses or so. You'd transition well before we arrived."

"But if I stayed awake for the remainder of my natural life span, I could direct the adjustments for another twenty-five or thirty ellipses at least. That direction would more than account for the disadvantages created by sim-Jack. I've run those numbers too." Lin sensed a hint of resignation in Eve's voice. Min was gone. She had no true motivation for reaching Uma. This way would at least maximize her contribution.

"I can't let you do that," said Lin finally. "And since it's my project, I get to decide how the analyses are done. So, no. You are going into repose when I do."

"I need to be helpful," protested Eve. "What can I do once we get to Uma?" She swallowed hard as her body began to tremble. "At least out here, I can still sort of pretend, you know? Once I see Uma, and the moon, it will be more real than I will be able to handle. And......" she added with a hoarse whisper, ".......I'm scared." She covered her face with her hand as Lin moved to comfort her.

"You are being helpful," Lin said quietly. "And I may well need you once we arrive. What if we are close but still require more adjustments? Who is going to help me with them? Jack or his sim? Hardly. Even though Min is dead, we can still save Sol. And we can save who knows how many other Umae who may need to venture back out into space. Please, I'll be there with you."

"So will I!" It was Jack. He was standing near the Life Lab entryway. It wasn't apparent how long he had been listening. He approached and sat down with a knowing smirk lighting up his face.

"What are you so happy about?" asked Lin without enthusiasm.

"I'm torn. I know something that you don't, but I can't decide if it will make me happier to tell you or just to keep it to myself."

"Jack stop!" snapped Eve. "Stop this! We are supposed to be helping each other for the Umae! You can't just keep secrets."

"Oh tell me about your secrets," said Jack with a sneer. "I'm that far gone, am I? DSM? Is that the diagnosis?" he added with a glance at Lin. "Jealousy. You two are jealous because you can't remotely approach my intellect and accomplishments. I refit the *Aurora*. I have the plan to defeat Dom and restore Uma. All the two of you have done is concoct some vague plan to try and cure a disease that no one really cares about."

"I suppose you are just going to kill him?" asked Lin.

"That's a simplistic summation, but yes. That's the general plan."

"You can't DO that, Jack. The Protocols forbid it."

"That's nonsense," exclaimed Jack.

"We are Umae!" protested Lin. "We don't kill!"

"If we don't kill, we won't be ANYTHING," countered Jack. "I, for one, plan on being great. The two of you? Well, you've tried to trick me but you didn't do a very good job."

"What are you talking about?" asked Lin. "Trick you? How?"

Jack slowly shook his head. "Stars you two are stupid. I know *everything*. You two have pathetically attempted to hide your grief over Min's death because you thought I wouldn't let either of you return to Uma if I knew he was dead. But I have checked everything. I leave nothing to chance. There is only one reason why you are both on this ship."

Eve's cheeks flared. "Well since you are so smart, why don't you explain that to us?"

"I'm smart enough to know that Min is still alive."

Eve steadied herself against Lin.

"Impossible," said Lin. "His life signs signature stopped transmitting a long time ago. Now you are just being cruel."

"No, the **Starshine** stopped *receiving* his life signs signature," said Jack slowly. "It was interrupted by the movement of the beacon satellites around Uma. But it's back now." He stared at Eve eager to take in her response.

"Eve, what happened after the signature stopped registering?" asked Lin.

Eve drew a deep breath. "I couldn't bear the silence, so I turned off the register. I thought......"

"That was your problem, as always," said Jack. "Come here." He stood up and walked to the Life Lab's Alpha panel followed by the two women. He manipulated its directives to relay the readings from Uma's primary tether beam. "Now watch."

Eve and Lin leaned in and anxiously focused on the display. The general data stream from Uma was essentially void of any meaningful information. But there was one regular package of data that repeated steadily.

"Oh......my stars," said Eve as tears began rolling down her cheeks. "He's alive."

"How long have you known this?" asked Lin.

Jack shrugged. "I've always known. Why do you think I agreed to let you come back to Uma on the **Aurora**, Lin? You have no role in my plans and my plans are complete. But showing Min......now that interests me a great deal.

165

So, here you are. And, it gave me a chance to see where your loyalties are. Not with me, certainly. So, let me give you some advice. Don't stop being useful."

Lin turned back towards the display.

"I'll do what you told me to do earlier," said Eve to Lin. "Now....I have to."

Lin rubbed Eve gently on the shoulder. "I know. And everyone will be better for it."

CHAPTER 16

The Umae looked on with confusion as the Hek walked off deeper into the forest. All of the Hek were going, and the Umae with Hek mates went with them. Left behind were the tribe's remaining Umae along with a scattering of Hybrids. They stayed beneath the cover of the forest canopy as the Sky Fire slowly descended below the eveningside horizon. The Umae kept their weapons close by as darkness surrounded them.

"Come, we must go!" implored Dom. He moved through the Umae, urging them to rise. "Now!"

The Umae muttered amongst themselves, their earlier confusion heightened. "Where are we going?" asked Joba. "I thought we were staying here."

"No," said Dom. "That was what I told them, but only because we need every advantage."

A buzz washed over the Umae. "What are you talking about?" asked one.

Dom waited for the group to settle. "They will return, but not as they say. I overheard their plan. When they return, they plan on killing us. Their excuse, that they need to protect Racha, is a ruse. They have left to form a plan of attack."

The previous buzz paled in comparison to the rumble of protest that shook the Umae. "Impossible! They are members of our tribe!"

"No!" bellowed Dom. "YOU are members of THEIR tribe." He paused to give the Umae a chance to consider his comment. "Recall how you do everything their way. Somehow they know what I told you and they know that you will no longer serve them. When they return they will seek to subjugate you, with blood."

"But they have protected us," protested a voice from the crowd. "Before you came here."

Dom pointed towards the night sky where a half-moon hung behind the leaves. "You have seen the Night Face? A message left by us after the Great Rain. We, the Journeyers, will not abandon you. And every Journeyer has been an Umae. There has never been a Journeyer who was a Hek. Whose protection do you value more, theirs or that of the Journeyers?"

The Umae began talking amongst themselves, many of them looking upwards and pointing at the moon. "The Night Face is Umae," said one finally. "It is like us."

Dom nodded. "Yes. Now you begin to see. And the Journeyers are like you.

167

Not the Hek. For ellipses beyond number, you feared them. And rightly so. You feared them because the Chroniclers told you that you should. And we, the Journeyers, told the Chroniclers why you should fear them. Trust me," he implored in a slightly quieter voice. "They will come back. And when they do, we must be ready."

"They are powerful," said another. "What can we do?"

"They are powerful," conceded Dom. "But we are more numerous. And we have our young that we must protect." As if on cue, an Umae child cried out. "And we will draw strength from that."

The Umae mothers in the group pulled their children closer. "You will lead us?" asked one.

Dom savored the moment. "Yes," he said slowly. "Of course. But we must go. We need time to prepare or all will be lost."

Without further request, the Umae began gathering their belongings. The Umae warriors checked the tips of their spears and scoured the ground for sling stones. Once the group was ready and had prepared their young for travel, they waited for additional guidance from Dom.

Dom pushed the Umae throughout the night. Pleas for rest were either ignored or granted with the briefest of respites. The Umae ate and drank on the move. Stragglers were encouraged and, in some instances, abandoned. They entered an area unfamiliar to most of them. The woods gave way to rocky ground scattered with short, scrubby bushes. By the time the Sky Fire began to appear in the morningside sky, all but Dom were at or near exhaustion. Just ahead was a steep, rocky hill leading to a plateau, a mesa rising upwards.

"Here is our salvation," said Dom. "We are nearly there." Without hesitation he continued forward and the Umae strove to stay with him.

The entire group paused at the base of the steep rise leading to the mesa. The top seemed to hang from the wispy clouds themselves as their fatigue clouded their estimates of the distance remaining. The rise itself did not seem as steep as it had on their approach. It would be a challenging climb with their equipment and their young under normal circumstances. In their present condition, with their aching feet and exhausted bodies, the climb would be extremely taxing.

"We are almost there!" implored Dom. His endurance seemed limitless. "Once we are at the top we can rest. But not until then!"

As he began the ascent the others lifted themselves from the ground, reclaimed their tools and their children, and started up after him. The rocky ground proved to be a benefit. The short outcroppings gave them sure footing that the brown grass would not. They pressed upwards, one struggling step at a time, urged forward by Dom's constant voice and their own longing for rest.

Dom would advance and then descend back to the group to shout encouragement. He would find those with failed legs and help lift them back to their feet. Finally, and only after the Sky Fire was well above the horizon, they made it to the top.

The top of the mesa was flat and mostly featureless. A few large rocks were scattered about but the shrubs they had seen below were mostly absent. At the far end of the mesa was a gathering of tall rocks. They leaned together forming a towering wall of stone. Dom summoned a small group of prominent Umae and led them to the far side of the structure. There was an opening that led downwards into the dark. A cool breeze greeted them from within.

Dom lit a torch and waved them forward. "Look!" They carefully followed him into the opening which was only wide enough for one Umae to pass at a time. The path on the other side of the opening led gently downwards into the shadows. The Umae followed Dom and his light, curious to see what he would show them.

The torchlight returned to them from the ground in shimmering sparkles. As Dom took the last few steps forward, they could see a spring-fed pool. They knelt down and drank as Dom looked on. They briefly bathed their battered feet, relishing the relief brought by the chilly water.

"You knew this was here?" asked one.

"No," said Dom. "My fellow Journeyers led me here. Now I, we, share it with you. Bring the rest up in small groups. Let them drink and bathe. The Hek will likely be here soon." He paused to allow this warning to sink in. "But fear not. The Journeyers have given me their guidance to share. We shall be rid of the Hek soon enough."

The Umae took turns availing themselves of the water in the cave. Dom made repeated trips up and down the face of the rise to drag exhausted Umae to the top. He didn't seem to sweat or to draw a labored breath regardless of how many he brought up with him. As the Sky Fire climbed higher it became apparent that the day would be unusually warm. Dom had as many of the Umae as he could rest inside the cool darkness of the cave, rotating them through for maximum benefit to the group. As they rested, he explained his plan.

"They will come after us," he said flatly. "There can be no doubt. And when they do, they will intend to do violence upon us. Now that they understand that you have seen through their deceit, you are of no use to them. They will slaughter us all if they can, maybe even to add us to their dinner fires."

The Umae gasped with disbelief each time Dom recited his message to a new group. But everything he had told them so far had proven to be true. He was a Journeyer, *their* Journeyer, and they would follow his guidance.

"What shall we do?" asked a large hunter. "You have brought us here for a reason. I'm sure of it."

Dom nodded. "You are wise. How are you called?"

"I am Shun," said the man.

"Well, Shun. I'm sure you can appreciate the advantages offered by high ground. We hold the highest ground in sight. And it will prove to be very beneficial to us when they come."

"We will have to fight?" asked a woman sitting next to Shun. Her face was slightly reddened from exposure to the Sky Fire and she had a light sprinkling of freckles on her nose.

"Hopefully for the last time," said Dom, attempting to sound comforting. "The plan I have should bring an end to their threat."

"Tell us," said Shun. "I have taught many of our hunters to use the spear and the sling. We can use them to protect ourselves."

Dom grinded his teeth briefly. "Indeed. I will share my plan with you, and you will share it with them. Then I wish to divide our hunters up into groups so we can be as organized as possible. This is what I have in mind."

The Umae all gathered in to better hear Dom's plan. Shun was particularly focused on Dom's strategy.

"I expect them to come charging up the hill that we just climbed," said Dom. "That is our advantage. I wanted you all to climb that hill while you were still tired so you would know how difficult it was. We will force them to do the same, but by then we will have restored ourselves."

"So we are to engage them as they reach the top?" asked Shun.

Dom shook his head. Perhaps Shun was not the best choice to place in a leadership position.

"No, as they are halfway to the top. Once they are halfway up, that is when we will strike. I suspect they will come after the Sky Fire has fled, so they will be using torchlight to see. We can remain at the top of the hill without torches. We will be able to see them, but they will have difficulties seeing us. Between that advantage and our higher ground, we will be victorious."

"But how do you know they will come in the dark?" asked the woman.

Dom fought back a smile. "Trust me. I know our enemy. I have been studying them for more ellipses than you have been on Uma. But walk with me for a moment, uhhhhh....."

"Arl. My name is Arl."

"Come with me Arl, I have a task that I would like for you to do."

The woman flashed a look of surprise at her fellows before rising and following Dom outside into the heat of the rotation.

"What would you like for me to do?" she asked nervously. "I'm not a warrior."

Dom smiled as he leaned in to whisper in her ear. "No, but you will do exactly what I say, won't you?" His voice was like a snake slithering through a tight space as a heavy, dark cloud flowed into her mind. She blinked hard a couple of times in an attempt to shake her confusion.

"Y-yes, of course."

"Of course, *My Lord*," added Dom.

"O-of course, My Lord," repeated Arl.

"Perfect," hissed Dom. "This is what I want you to do. And let no one see you leave."

Arl nodded absently as Dom leaned in once more and gave her his directives.

Dom waited until the Sky Fire was low in the eveningside sky before moving to the top of the hill with Arl. The majority of the other Umae were still refreshing themselves with the water and shade provided by the cave. Dom was sure to command Shun and his hunters to be well-rested in case of attack. He had directed them all inside the cave, assuring them that he would maintain the necessary watch.

Arl stood facing morningside. Her expression was distant as she stared out over the expanse of ground below them. Dom stood next to her, focusing his attention on the horizon until the last light of the rotation began to wane.

"Yes," he muttered. "There they are."

Arl squinted. "There who are?" she asked dreamily. "I don't see anyone."

Dom didn't doubt that. His senses were far keener than these pitiable Umae. His eyes had already detected the tell-tale flicker of torchlight moving in their direction. It would be some time yet before Arl's eyes would be able to see it.

"You will," said Dom. "Kneel down." Once she complied, he took her chin in his hand and redirected her eyes upwards. A single bright star dominated the sky in that direction. "Start walking," he explained. "And use that star to guide your way. Even though you can't see them yet, if you follow that star, you will see them soon. And once you do, you will do as I instructed."

"Y-yes," replied Arl. "What you said."

"You do remember, don't you?" asked Dom. He had attempted to temper his degree of control so as to leave the Umae woman with enough of her cognitive abilities that she could follow a set of directives. As complete domination had always been his goal, he had never attempted this before. He was confident, as always, that he had achieved his goal.

"Yes," said Arl. Her voice reflected a deeper confidence.

"Then go," said Dom.

The woman rose and started slowly down the face of the steep hill. She was careful to turn sideways and utilize the scattered rocks as foot holds so she wouldn't slip. This encouraged Dom. She hadn't lost all of her ability to think. Now if her memory would only prove as serviceable.

Arl made it to the bottom without incident. Glancing up at the sky, she headed away from the base of the hill. Dom returned to the cave to check on Shun and his men.

She began walking morningside. Once the last of the Sky Fire's light was gone she stopped briefly to ignite a torch. Dom didn't care if she found them or if they found her. The result would be the same.

She walked briskly through a wooded area. The effective radius of her torch was limited by the compact arrangement of the trees nearby. They all cast shadows that stretched out behind them at varying angles as she went by. It occurred to her that being out in the forest, all alone at night, wasn't the safest situation for her to be in. However, despite the recognition of this danger, she felt no fear. Dom had given her directions and she was going to follow them.

Her feet began to grow tired. She occasionally heard the call of various unknown creatures from beyond the range of her light. Several times she had to maneuver so she could see through the canopy well enough to locate her guiding star. She also had to stop and light a new torch from the old one as it consumed itself. The noises of the strange animals seemed to be growing louder. Still, she knew no fear.

The trees finally gave way to a broad, flat plain. The grasses and shrubs rose up to nearly her waist and she knew that this was a prime hunting ground for predators. The plant life rustled as she waded through it, holding her torch up high away from the grass. The ground rose up to a low plateau. Once she reached the top, she stopped to gather her bearings. Her star was still directly in front of her. As she tried to study the terrain ahead of her, she finally saw what she had been waiting for. A line of tiny lights snaked its way across the plain in her direction. The Hek were coming. Now she only had to remember what Dom had told her to do.

Shun and his fellow Umae were restless. They sat inside the cave, sharpening their spears by the firelight. A large pile of sling stones had been stacked by the entrance. All of these developments were pursuant to Dom's plans.

"I feel it will be soon," said Dom somberly. "Before the Sky Fire appears again."

"But why?" asked one of the warriors. "Why do we attack them?"

Dom glared at the man briefly, causing the hunter to avert his eyes. "We do not," said Dom. "We simply protect ourselves. They will come for us. You must trust me. The Hek are treacherous. They will seek our deaths, but we shall defeat them instead."

"What of the others, those outside the cave?" asked Shun. "What role will they play?"

Dom resisted the urge to chuckle. "Ah yes. The women. I have given them the appropriate instructions. They will not fight, but they will not be idle either. Everyone, save the youngest of the children, will contribute to our defense."

"This hill, it is good ground," noted Shun. "The daggercats use such places to see far away. We can do the same, at least when the Sky Fire soars."

"And the Sky Fire shall not soar again until we have defeated them," noted Dom. "As I said before, they will be here soon."

"But how will we know when they are coming?" asked Shun. His voice was weighted with a concern that his fellows plainly shared. "If we use torches to light up the hillside, it will simply make it easier for them to find us."

Dom nodded. "You are a wise leader," he said enthusiastically. "But we need not worry about the Hek using stealth in their attack. They are as bold and overconfident as they are treacherous. We will see them coming. We will hear them coming. And we will do that because they will want us to."

"I know them," said another warrior. "They are brave, but not foolish. When we hunt the great bear, they make careful plans. They will do so now."

"No," said Dom coldly. "They do not respect us as they do the bear. This is a celebration for them. It is a slaughter they welcome as a successful hunter welcomes the first cut of meat from his fire. They know no fear of us. But that will be their undoing."

Shun nodded. He made a visual survey of his fellows. He knew them well.

They would follow Dom and protect their fellow Umae. "So tell us, Great Leader. What is your plan?"

Dom gripped Shun by the shoulder. "Each group has its own role," he explained. "I've already advised the women and children as to what their actions will be. As for all of you," he spread his arms wide, "you shall be our heroes. When the Hek attack, we will be positioned and ready for them. So," he said, turning towards the mouth of the cave, "sharpen your spears and ready your slings. Shun, divide your men into two groups. One group shall be those who favor the spear, the other the sling. Stack your extra spears and stones along the edge of the plateau on the morningside side of the hill. That is where we shall find our victory."

Arl ran as fast as she could through the grass. The lights in the distance seemed to turn and move in her direction. They must have seen her. Her breath shortened and she finally had to stop. Her torch flickered and died. She stood in the darkness gasping for air.

As the lights approached her, she better appreciated their number. From a distance it had been hard for her to make an estimate as to the size of the group. There were far more than she had expected. As the light of the Heks' torches enabled her to see them up close, she realized that there were at least as many of them as there were Umae atop the hill. They were nearly naked and bore thin spears. Their bodies glistened in the light of their torches. Despite the fact that they had just sprinted across a wide, grassy plain, none of them seemed to be breathing heavily. Even the female Hek warriors, although smaller and not quite as muscular as their male counterparts, looked as if they had just spent the day resting beneath a shady tree. Now, she was supposed to say something......?

"You talk," said one of the Hek as he handed her a skin full of water. "No safe here. Why you come?"

She accepted the skin and took a long pull from it. Her breath was returning.

"It is......bad," she said. She had never communicated very much with a Hek. She couldn't tell if they understood her.

"What bad?" said another Hek man as he stepped forward. Arl recognized him as Palu. "Where clan?"

"Clan?" she echoed. "The Umae?" Her ears rang as her thoughts bounded around inside her head.

The Hek warrior nodded. "Umae."

"They need help," she managed finally. "Sent me to find you. Knew you would come."

"Help what?" said Palu urgently. The other Hek were becoming agitated.

Arl closed her eyes and tried to remember. Dom's words swept her own thoughts from her mind. "A group of Umae. Not ours. Bad. They are attacking. So many. Chase our people away."

Various members of the Hek group began howling when they heard this. Palu raised his hands to quiet them. "Where? Where they chase?"

Arl put her hands over her ears, trying to keep Dom's directives from escaping. "Dom said they would seek a large hill, with water. He said you would know. Said your lights would scare them off."

Palu turned and faced the other Hek warriors. He began barking orders in Hek that Arl could not understand. Those who were not bearing torches quickly took one in hand and lit it from a fellow's. The entire Hek group blazed with light.

"Know place," said Palu. He took a nearby warrior by the arm and pulled him forward. "Croja keep safe. You stay with him. We go and help Umae." Palu waved to the others and barked several more directives. The group took off in a gleaming wave across the grassy field leaving Arl and Croja behind.

Arl absently took a couple of steps closer to the muscle bound Hek warrior. For the first time since she had left the hilltop, she was afraid.

Torches blazing, Palu led the Hek across the grassy field towards the plateau where Arl had told him the Umae were under attack. The top of the plateau was dark, but as his Hek warriors halted to take direction from him they could hear the sounds of weapons clashing and women crying out. Palu made a quick survey of the steep climb ahead of them. "Upwards!" he implored. "They are in trouble!" He waved his warriors forward.

They spread out in a wide line, each one eager to be the first to reach the top of the plateau. Although their torches provided them with plenty of light, the terrain was still treacherously steep and covered with rocks. The warriors leaned forwards into the slope and drove themselves upwards with their powerful legs.

Just as the last of the Hek began their ascent, they came under bombardment by a hail of spears. With no cover and their ability to maneuver compromised by the slope, the Hek could do little to defend themselves from the attack. Many were impaled while others were knocked off-balance before toppling backwards down the slope.

"Courage!" urged Palu. "Keep moving!" Just as the Hek leader pressed forward, a spear spiked into his stomach dropping him to his side. He clutched at it desperately with both hands as his blood began to soak the side of the plateau. His veterans understood their obligations. One attended their fallen leader while the others renewed the charge up the hillside.

The Hek looked upwards in an attempt to see their attackers. They could only see shadowy figures as the top of the plateau was still some distance away and was completely dark. Their own torchlight could not reach far enough above them to enable a view of the summit. The sounds of weapons clashing and women crying out still summoned them forward. The Umae were in desperate straits.

Despite the slope and the rocks that defied them to get a decent foothold, the Hek redoubled their efforts. Those who had endured the spear assault forced themselves higher. As they did, a second wave of spears poured down on them. Still, they could see very little of what was ahead of them and the slope made it impossible for them to hurl their own spears far enough to reach the top. The Hek joined together in a deep, rumbling war chant as they sought to bind their spirits together. Once again, they denied their injuries and the rise before them and drove themselves towards their attackers.

Just as they seemed to make progress, they were met with a torrent of sling stones. Hek after Hek grunted in pain as the stones pelted them, spilling even more of their blood. Some suffered bone-breaking attacks as the stones struck their heads and arms as they tried to cover themselves. The Hek's charge was halted as they fell backwards from this assault. They lay down on the face of the hill to make themselves harder targets.

"Our torches aren't helping!" cried one. "We can't see far enough ahead and they slow us down!" As if on cue, his fellows began tossing their torches aside making both arms available to them for climbing. With a loud grunt from this new leader, they rose to make one last effort at the summit.

Almost all of them carried the burden of injury. Some tried to push past the pain of a spear wound while others were bleeding badly from the impact of so many sling stones. Blood loss was exhausting their energies. They still could not see their attackers clearly. Many were using the butts of their spears as crutches to push themselves along.

As they finally neared the top, they could see who it was that had caused them so much pain. A large group of familiar Umae launched a final attack of spears and sling stones dropping all but a handful of the Hek. The few who did make it to the top could barely stand. They looked around the plateau in disbelief. Behind a line of spear-bearing Umae were a group of women banging spears together and crying out. There were no fallen bodies. There were no signs of any type of attack.

The new leader dropped his spear and held out his arms. "Why you......?" His inquiry was cut short as a spear pierced his throat. His blood shot upwards like a fountain as he collapsed. Dom, having directed the spear, glared at the other Hek.

"Kill them," he growled. "Kill them all."

The Umae surged forward with their spears. The Hek were barely able to defend themselves. A few attempted to flee back down the slope only to be struck down by spears. The remainder fought with bravery, but their broken beaten bodies could not respond. The remaining Hek were assaulted again and again by the raging Umae until none of them moved.

The Umae then moved methodically down the slope with their spears, putting down any Hek who persevered. Soon, the Hek were dead. All of them. The Umae had suffered only two casualties.

Dom stood at the top of the rise, watching the carnage. One of the Hek who had fallen nearby let out a quiet groan. He was met with Dom's foot on his throat. While this was just the beginning, Dom knew the Umae belonged to him now to do with as he chose. And he was going to make them pay for what the Journeyers had done to him.

Chapter 17

Lin and Eve watched as the last piece of a visual puzzle slid into place. Eve had transformed the variables in Lin's RTS research into geometric patterns on the Life Lab's main display. It helped Lin more intuitively understand how each of the factors interacted and weighed on one another. The visual renderings allowed her to make observations and insights about the nature of her problem that she would not have otherwise made using conventional mathematics. Lin was almost giddy once the final steps were completed.

"Eve, I really think this is going to work!" she gushed. "I don't think anyone would have ever thought to attempt the type of compounds we discovered during our analysis of the aliens. It's amazing. Nature provided them with an answer to this problem through evolution and we couldn't figure it out even though we knew what we were looking for."

For the first time in a long time, Lin saw Eve smile. "That's one of the things that makes Data Science so interesting," she said. "The way things are connected. Correlations distinguished from causations.....the universe willingly provides so many clues to so many mysteries. We just have to know how to look for them."

"There really isn't anything left for us to do," said Lin solemnly. "The studies are all planned and sim-Jack is going to monitor them for us while we are in repose."

"What about the real Jack?" asked Eve. "Should we check on him?"

From what the two women had seen, Jack had not emerged from his chamber for over sixty rotations. They had no way to communicate with him.

"It won't make any difference," said Lin. "He won't come out, or do anything for that matter, until he is ready. We should schedule our repose and be sure we are up again with plenty of time to review the results of those studies. We will need some manufacturing time, but based on those compounds, I don't think that should take very long."

"You sound very confident."

Lin smiled. "I am. This is it. I wouldn't say so if I weren't sure because I wouldn't want to get your hopes up."

"Then this is it," agreed Eve. "I'm going to get to see him again. It doesn't seem possible."

"Come on. It's time for us to go to sleep. Everything will be quite different when we awaken."

"I'm sure. Thank you for all of your work. Your work with me, with your studies. I owe you so much."

Lin gave Eve a warm hug. "I didn't do it entirely for you. Or for Min. Or even for Sol. The Umae need us, Eve. They always have and they always will. Their future will, sooner or later, lead them to leave Uma."

"And ours leads us back," said Eve.

"Indeed."

The two women left the Life Lab and walked to the Repose Center. After entering directives concerning the length of their sleep, they turned towards one another.

"Sweet dreams, Eve," said Lin.

"Oh, of that I have little doubt."

CHAPTER 18

Dom ordered everyone out of the cave atop the plateau. Low torch light sparkled off the surface of the spring-fed pool. He sat on a flat rock near the edge of the water awaiting the visitors he had arranged to meet. His followers had taken to referring to him as "Chikwa". It was a Hek word denoting great power and benevolence. Despite his hatred for the Hek, Dom allowed it as it conveyed the perception he wished them to have. This settlement became known as "High Hill". Dom approved. It was simple but implied conquest and broad perspective.

He didn't have to wait for very long. The Sky Fire had almost completely disappeared and a group of women slowly filed in from the outside. Dom had commanded Shun to make sure that no one interrupted this meeting or attempted to observe what was happening inside. Always dutiful, the Umae Battle Leader created a perimeter of his most trusted warriors around the cave's opening. There would be strong magic inside the cave this night and only Chikwa and those chosen to go inside would experience it.

A group of young women were permitted inside the perimeter and directed to enter into the cave. None of them had more than twenty ellipses but all had at least thirteen. As they entered, Dom bade them to be seated on the floor in front of him. Slightly in awe of his presence, they complied.

"How many of you had a Hek as a mate?" he asked.

Hesitantly, slightly more than half of them raised their hands.

"And how many of you have borne young with your Hek mate?"

All of their hands dropped as many of them lowered their eyes with apprehension.

Dom had chosen this group carefully. He knew that the members with more ellipses had all had Hek mates. He also knew that all of the younger ones had only just reached fertility and could not have yet had an opportunity to become pregnant. "As I have said. The Hek have long withheld their seed from the Umae women in our clan. It was a part of their dark plan. They sought to deny us our numbers while they secretly bred with their own. Eventually, they would have outnumbered us and we would have been at their mercy."

The women muttered anxiously amongst themselves.

"But they have been defeated haven't they, Chikwa?"

Dom nodded. "For now, but the battle we face is a long one. They will return again and again over time to test our numbers. This is why we meet.

I am going to share with you the secrets of reproduction. I will share it with every Umae woman able to bear young. And you will take these secrets to your mates. And if you do not have a mate, I will find one for you. To be strong, we must be many. To be many, you must do as I explain."

"But, what if we can't do it?" asked one of the older women. "Is it difficult? What if we are unable to learn?"

Dom stood up slowly and turned away from the group, unable to contain the broad smirk on his face.

"Have we displeased you, Chikwa?"

Dom drew a deep breath and faced them again. "No, of course not. You will have no problem remembering. Trust in my teaching."

"Yes. Yes, of course Chikwa."

Dom waved them forward. After they were entirely focused on his words, he spoke again, this time in a deep, soft voice that flowed into their heads like warm water.

"You will form a line," he said slowly, "and then remove your cover. As I remove mine." He stepped from the loose wrappings that hid his small body, allowing them to drop to the floor. The women were doing likewise, oblivious to the fact that he was now completely naked. Once all of the women were completely bare, they formed a line in front of him. "You will not speak of this with anyone," he droned. "You will not remember any of this. Once you leave this chamber, you will seek out your mates and you will couple with them. After that, you will couple with them as often as you can until you are with child. If you do not have a mate, you will seek one out by enticing him to lay down with you. You will tell the men that an important part of my plan is for them to join with you as frequently as possible. Do you understand?"

All of the women stared at him with glassy-eyed expressions. "Yes," they replied in a gentle unison. "We understand."

"Very well. Now, one at a time, approach me. Approach me and I will bestow upon you the greatest gift any Umae woman could ever receive."

Word of the great Chikwa spread across the land as tensions between the Hek and the Umae grew. His plateau was a place of sanctuary. Once a group of wayward Umae arrived, they were quickly assimilated into the clan. Women of child-bearing age underwent Dom's wisdom training and were sent back to their mates. Those without mates were encouraged to find one or one was provided. The men in the clan were entirely pleased by Dom's wisdom. As the size of the clan grew, the settlement spread out away from the plateau. The Hek did not dare approach this place as Dom explained that they held great fear of him and his power. More dwellings were built. More land was cultivated. Dom would not allow them to follow the ways of the Hek. The Umae would grow crops and raise their own animals.

By the time the night's face stared down at them again, every Umae woman in the clan within a certain age range was pregnant. Dom's wisdom was powerful indeed. And he ruled over them all from the half-light in the Chamber of Waters perched atop his rocky throne wondering about the glories to which he would lead all of his children. There would be other Umae groups who would not come. But they would be dealt with eventually. For now, Dom was quite content to share his wisdom with the Umae, knowing that the perversion of their bloodline would serve as both the tool for his ultimate conquest as well as the revenge he had sworn upon Min and his fellows.

As word of High Hill's grandeur spread, Umae from all over the area came and joined. Ground was tilled and planted under the instruction of the Umae who recalled life prior to the Great Rain. Animals were captured and enclosed. And those who had learned the art of the hunt from the Hek stalked the land with spears and slings and brought back wild game. As the shelters for the members of the community began to spread out away from the base of High Hill the children of Dom's wisdom were born.

They were strong, noisy infants. Their cries became so commonplace that they soon went mostly unnoticed by those few without children. The men in the settlement beamed with pride at their healthy offspring, not suspecting they had been cuckolded by the one they would have least suspected. The babies pulled relentlessly at their mothers' breasts as their growing bodies craved as much nutrition as they could consume.

Soon after a woman would give birth, she would be with child again. The men strutted around the settlement, very satisfied with their own perceived potency. And Dom ruled over them all, summoning the women to his cave periodically for more indoctrinations about his wisdom. The settlement spread out across the land like a patch of wild ivy. There was no structure, no plan. The shelters would go up wherever anyone decided to raise one. The lands were plowed and the animals managed through the advice of Dom by his proxies. He rarely came out from his cave. He had special ceremonies that allowed his people to come to him. The women were granted his wisdom. The men were reassured of their own virility and power. And all of those who left Dom's presence were filled with the certitude that, because of him, their future was guaranteed to be happy and successful.

As the ellipses passed, fewer and fewer sightings of the Hek occurred. The hunting grounds of High Hill were far and wide, and the Hek had seemingly disappeared. Any Umae who were encountered from outside the settlement were quickly assimilated. They had no choice. Not because of any threat of force, but because of the obvious vitality and success of Dom's people. And while they had no Hek with them, they did bring the Hybrids. They were tall and well-built, although not as stout as their Hek kindred. Their skin tanned against the unrelenting Sky Fire. They learned quickly and were curious to a fault. Their innovative natures led them to ask many questions and to improve upon the farming and husbandry methods employed by their Umae companions.

One evening just after the departure of the Sky Fire, Dom summoned a group of women to his cave for a lesson in wisdom. They were mostly new arrivals who had not yet met him in person. They were still hushed by a sense of awe having heard stories about all of the great things he had done for his people. Anxiety and curiosity swirled within their breasts as they entered the cave to take audience with Chikwa.

As was his way, Dom sat silently on a large throne of stone he had directed the Umae stone smiths to carve for him. The women passed slowly by stealing glances at him in the low light. Once they were inside and the entrance was secured he rose and beckoned to them.

"Welcome!" he said in a grand voice. He allowed the sound to resonate. "Welcome," he added in a much quieter, darker tone. "I bring you wisdom."

He surveyed the group. There were sixteen Umae women and, for the first time, two Hybrids in this group. He had never had a Hybrid as his student.

"You there," he said, pointing to a Hybrid girl near the front. "How are you called?"

"I am Weena," she answered nervously. Her stomach fluttered at the prospect of being singled out by the forceful child seated on the stone throne.

"'Weena," repeated Dom as he carefully looked her over. "You are new here."

Weena nodded. "Yes. We arrived the rotation before last. Thank you for welcoming us."

Dom grinned. "Your welcome is incomplete," he said softly. "But I will grant you wisdom." He climbed down from his throne and approached the group. "I will grant all of you wisdom." He once again employed the same low, deep whisper. The women became focused on his voice. "And you shall bear mighty offspring. But you will remember none of what I have taught you. You will remember none of what transpires here."

"We will not remember," droned the women.

"I-m sorry," said Weena feebly, "but I don't understand. How can you impart wisdom upon us if we won't remember anything you teach us?"

Dom's back stiffened. His face flashed red hot as the blood rushed to his face. "You.....ask questions?" he spat.

Weena trembled. "I'm just trying to understand."

"Come here," said Dom. His voice was like a new fresh breeze of cool air. His focus was entirely on Weena.

"I'm afraid," Weena admitted.

Dom loosed a low growl. "You refuse me?" His question was laden with confusion.

"What do you want me to do?" she asked.

"What I want," he began with a failing restraint, "is for you to come....... here."

Weena looked at the other women. With the exception of the other Hybrid, they were all standing transfixed on Dom's figure. Timidly, she made her way forward to where Dom stood and knelt down before him, unsure as to what was expected.

"Very well," he said. "Now listen." He took her face in his hands and peered into her tear-filled eyes. His voice was now a smooth flowing shadow. "You want to serve me, don't you?"

A tear trickled down her cheek. "Serve you? I don't understand. You are the leader. But don't we serve each other?"

Dom's anger overcame his restraint. Weena's skull sounded like the slow breaking of a dry branch as Dom gripped her head. He released her and her corpse toppled awkwardly to the ground.

"Noooooo!" It was the other Hybrid woman. She turned and sprinted towards the cave's opening before disappearing outside. The other women did not budge as Weena's blood spilled out onto the chamber floor.

"Yes," growled Dom in a deep, low growl. "Each of you. Come forward one at a time. I will bestow upon you the wisdom. You will forget what has happened here." The women lined up and waited in turn to approach him.

Dom glared at the dead body bleeding on his floor. "You will forget. But I will not."

Dom brooded for a long time. Rotations passed and no one in the High Hill community saw him at all. Upon strict orders, no one was to approach the Chamber of Waters. No other directions were provided. No wisdom was granted. His people eventually had to make plans to retrieve water from nearby sources as they had no access to the spring in the cave.

Finally, he emerged. Without a word he strode boldly down the slope to the collection of dwellings below and let it be known that he wanted every Hybrid claiming membership in his clan to follow him. Their number was relatively small, perhaps no more than 20. The girl who fled following Weena's murder was not among them. No one, it seemed, knew what had become of her. A pair of Umae guards had seen her run from the cave and down the slope but had dared not abandon their posts or interrupt what was happening inside. She apparently had no kin and had chosen to flee into the woods in the dark of night. The Umae assured themselves that she must have gone mad.

Dom never asked about her again. Instead, he gathered the Hybrids together and told them a story.

"You are the children of both the Umae and the Hek," he explained. "Your blessings are bountiful. You are clever and innovative like the Umae. And you are strong and brave like the Hek, although without their dark hearts. These gifts make you special and you shall have a special place by my side."

A few of the Hybrids still struggled with the Umae language, but there were those among them who could translate Dom's words into those of the Hek. The Hybrids beamed at Dom's high opinion of them.

"What would you have us do?" asked a male. He had the most ellipses of any in the group and beamed with confidence. Dom regarded him as a curiosity.

"Tell me, what is your story?" asked Dom. His interest seemed sincere. "Where do you come from and how did you get here?"

"I am Vankin," said the man. "I have traveled a long, long time. I grew up on a high mountain plain many, many rotations from here. My father was a gift to my community, offered by a settlement of Umae. He and my mother produced many offspring."

"Where are your siblings?" asked Dom.

Vankin drew a deep breath and attempted to maintain his bearing. "Lost. When the Great Rain came they were swept away. Our plain was encircled by high hills and mountains. The rains turned our land into a vast lake. There were few ways to escape and most of my people perished."

"Your people, how many?"

Vankin thought for a moment. "I would say a few hundred before the Rain. Most were Hek, but many were Umae. A few were like me."

"And after the Rain, where did you go?"

"I had managed to find a high point to keep myself alive. But it was very cold, at least until the Sky Fire emerged. I found only two others from my community who had survived, so we made our way down to the base of the mountains. The water was everywhere. We could not go to the places we had gone before. They were all gone."

"Continue."

"We walked along the high ground towards the eveningside, traveling at night. The Sky Fire burned our skin at first, so we hid during the day. I think the rains must have frightened away any dangerous creatures. We had our spears and our courage, but we would not have been able to defeat a daggercat if one had found us."

"And now you are here."

Vankin nodded. "Yes, but it took a long, long time. We eventually met up with others – Hek, Umae and those like me. Our wandering band grew larger. At first, everyone helped everyone else. But as our group grew larger, there were more disagreements. We began to fight with one another. Eventually, we were attacked by another group and much blood was spilled."

"Hek?"

The Hybrid hesitated. "Not entirely. They were a group like ours, a mixture of the three. But they had planted crops and were not willing to help us with food. They saw us as thieves, I think. Our numbers were greatly reduced so we fled."

"How long ago was that?"

"I have been here for just a few rotations," said Vankin. "After our battle, I would say that we walked for another fifty rotations before we got here. We wanted to make sure that we were far enough away from those who had attacked us."

"And how many were in your group?"

"Four. Four out of what was at one time close to thirty. Myself and three Umae."

Dom closed his eyes and considered Vankin's story. "You know now that the Hek are the source of all conflict?"

Vankin nodded. "That is what I have been told. I wouldn't have believed it because the Hek in my group were not like that."

"Tell me, did your settlement raise its own crops and keep its own animals?"

"No. It was too chilly. We had trails we used to get down the mountains to search for food."

"And animals?"

"The Hek in our tribe believed it was easier to hunt than it was to raise our own, so that is what they did."

"And who did this hunting?" asked Dom.

"Well, the Hek did. Mostly."

Dom smiled. "So the Hek controlled your group's food supply?"

By now Vankin was fidgeting nervously. "Yes. I guess that's right."

"And tell me this – did the Umae in your community know how to grow crops?"

"There were those who spoke of raising crops in the days long before the Great Rain," said Vankin.

"And who was it that chose the location of that community, the one on the high mountain plain?"

"The Hek said it was a good place to hunt," said Vankin. "But I don't know who chose it. It was before my time."

Dom stepped back and addressed the entire group. "I have heard story after story like Vankin's," he said. "The Hek made the decisions and the Umae simply went along. It was the intent of the Hek to enslave the Umae. They determined long ago that the Umae were beneath them. Vankin, how many offspring have you seen born to a Hek parent and an Umae parent since the Great Rains? You have traveled far and wide......."

The man shrugged. "None. Not a single one. All of the offspring were born to either Umae parents or parents who were both like me."

Dom released a growl so deep and threatening that it made the entire group scramble back in fear. "The Hybrids are breeding with each other?" he screeched.

Vankin swallowed. "'Hybrids?' You mean, those like me?"

"Yes! Those like you!"

He nodded. "Well, of course. Why wouldn't they?"

Dom's entire body started shaking. He paced back and forth like an anxious mother bear. "At the appearance of the Sky Fire next rotation, gather here again!" he demanded. "And prepare for a journey." He drew a long, slow breath to gather himself. "It will be quite........important."

Before the Sky Fire rose above the morningside horizon, Dom was stalking from shelter to shelter at the base of the rise. He was searching for the

Hybrids who were going with him on the promised journey. A very important journey.

Gradually, the Hybrids arrived and reluctantly accepted the spears they were offered by Dom's Umae. Most of them looked rather uncomfortable with the weapons, having never needed to employ one. A few handled them with confidence, feeling more assured that they would have some protection for their trip to....... Dom had not been specific. He had promised it would be "important", but that was all. The spear handlers didn't seem to care. Their companions clustered closely together, their anxiety almost palpable.

Dom counted and recounted until the final Hybrid arrived. The Sky Fire had lifted the darkness, revealing a dour expression on his face. The tone of his voice was no less severe.

"We must be swift," he growled. He glanced over at the nearby Umae. "Leave us." The Umae slowly wandered off, their curiosity about this gathering unsatisfied. With a gesture, Dom beckoned the Hybrids to follow him.

He led the group past the far boundary of the community that had grown up around Dom's plateau. He abruptly stopped and turned to address them. "There is a task that must be accomplished, and it can only be accomplished with you. It is vital to the success of my plan.....my plan for this community."

"What is it that you would have us do?" asked Vankin. "You have given us arms, but I do not think that many among us are warriors. Is this not something better suited for the Umae?"

Dom's eyes narrowed. "You are warriors," he said plainly. "Your Hekish heritage assures that. The fact that you have not yet had to act as such means nothing. All will be revealed."

Still confused, Vankin looked at his fellows to determine if any of them had a hint of enlightenment. It appeared not as the others were also searching for insight. With few other options, they remained silent and followed Dom as he started off again heading farther away from the settlement.

The Sky Fire was particularly cruel that rotation. The wispy white clouds that typically floated above them were absent. The wind was completely still. As they walked along they could hear only the sounds of their footfalls in the scrub grass beneath their feet and the buzz of an occasional bug stirred by their passage.

They had brought no water. The Hybrids did not know any details of their mission and had not known how to prepare. The Umae had not provided them with water so the Hybrids assumed none would be needed. They were wrong. As the searing overhead eye beamed down on them and sweat dripped from their faces, their bodies cried for hydration. Even those Hybrids who had spent enough time beneath the Sky Fire for their skin to darken slightly saw their skin grow red by mid-day. Dom's did not. His Umae skin retained its pale pallor. He did not sweat. His step did not falter in the heat.

"Dom," said Vankin finally. "We must stop! We require water. And shade. We are burning up!"

Dom reviewed Vankin's arms. The Hybrids had skin that was inherently darker than that of the Umae. But the Sky Fire was such that Vankin seemed to glow red hot. His companions were no better off. "There is a creek nearby," said Dom. "But it is frequented by many animals. And those animals are preyed upon by others. Are you certain you want to go there?"

Without hesitation, the Hybrids offered their consent. Dom regarded the Sky Fire for a moment as if to gather his bearings and then led them off in a new direction.

This new path led them through a wide plain of waist high grass. As the Hybrids passed, the grass rubbed against their skin causing an intense itching. Dom did not seem to notice. "Are we close?" asked Vankin as he dug at the back of his hand.

"We must continue." He moved forward at a brisker pace. The grasses left the Hybrids' skin covered in scratches from the rough edges of the plants as well as welts from whatever the grass was transferring to them. The Hybrids pressed on, eager for water.

The plain gave way to a forest of medium-sized trees. Although they did not provide any relief from the damage done by the sharp-bladed grasses, the trees at least served as a canopy to block the Sky Fire. It gave the Hybrids some measure of succor that pushed them forward. The water had to be near.

"Over that ridge," said Dom finally. By now the Sky Fire was in descent. The Hybrids' skin raged with the burn it left behind as well as the torments offered by the cruel grass on the plain. Water would bring them relief from all these maladies.

They made themselves move ahead up the ridge. Their spirits lifted as they reached its zenith so they could survey the ground ahead of them. As Dom had promised, there was a wide winding creek bed. But it was dry. Instead of cool, flowing water it offered nothing but dust and rocks. The Hybrids all looked at Dom with disbelief, hopeful that there had been some mistake.

"It appears to be dry," said Dom matter-of-factly. "A shame."

"A shame?!?" shrieked one of the Hybrids. It was a male, one of the spear-handlers. "We need water! You told us there would be water!"

189

Dom's mouth curled into a slow smile. "Perhaps I should tell you a story. Please, form a circle around me and I will explain."

Their minds buzzing with confusion, the Hybrids slowly encircled Dom and waited for him to continue. They knelt down, some leaning on their spears and others dropping them entirely. Although some of the Hybrids were heartier than others, they were all exhausted. The Sky Fire, the grass, and the lack of water had taken their toll. It felt good to rest.

Dom started laughing. It began as a mild giggle before blossoming into a roaring cackle. The Hybrids searched one another once more hoping that someone understood. No one did. As Vankin moved forward, Dom's affect abruptly changed. His laughter was completely gone, replaced by a murderous glower.

"Here's my story," he said finally. "I am a liar. I am a conqueror. I brought you here to kill you."

A few of the Hybrids squealed in fear but most of them began muttering to one another. A couple stood and defiantly seized their spears. "And just how will you do that?" said one. "There are at least twenty of us and we are armed. There is only one of you and you bear no weapons at all. And you are encircled. I think the Sky Fire has cooked your mind."

A few more were emboldened by this man's words. They also gripped their spears tightly and focused on Dom. "You are not so fortunate," said Dom. "Perhaps you should press your advantage?"

With a glance towards the man next to him, the Hybrid sprang forward aiming the point of his spear at Dom's head. Dom easily avoided the attack as well as the one directed by the second man. Both men toppled off balance as Dom maneuvered to utilize their momentum against them. With a short strike to each of their backs, the men fell to the ground. The other Hybrids, some more confident than others, slowly moved in. As they did, Dom produced a small piece of carved bone from his pocket. He held it to his mouth and blew.

A shrill whistle raced away from the scene causing the Hybrids to cover their ears in pain. The sound was so loud that it made them unsteady on their feet. Dom stood in the center of the circle, waiting.

As the Hybrids recovered their bearings, a large group of Umae burst from the trees. They rained sling stones down on the Hybrids as they charged, leaving several Hybrids prone and bloodied on the ground. The other Umae closed with their spears and engaged the wavering Hybrids, quickly cutting them down. The spear-handlers stood side by side but, regarding the odds, turned and attempted to retreat. The rotation spent in the Sky Fire and all of its attendant tortures had sapped their strength. The Umae easily caught up with them and they soon fell as well. The conflict was nasty, brutal, and short.

As the Umae made sure that all of the Hybrid were dead, Joba approached Dom. "You were right. The Hybrids are traitors."

Dom nodded. "Of course. It is there Hekish blood. They are tainted. But now we know for certain, don't we?"

"Yes, we know for sure. I will be sure and tell everyone back at High Hill. We must be careful in the future."

"Oh we will be," said Dom assuredly. "We will be."

The Umae stood at the bottom of the hill, howling threats towards the wooden palisade at the top. They eagerly banged their spears together driving their blood to a boil.

Joba stepped forward to the front. A red cape streamed from his shoulders and a crested helmet covered his head. He removed the helmet and tucked it under one arm, his old eyes peering upwards towards the enemy.

He had done this many times by now. The deep lines in his face were a testament to the many rotations he had stood beneath the Sky Fire and readied his troops to face the Hybrids in battle. But now his enemy was employing a new tactic. In earlier times, the Umae had simply scouted their position and then encircled or flanked them until the outcome of the battle was predetermined. But the Hybrids had grown wiser with each battle. They screened their positions with scouts making the old Umae tactics obsolete. Then they began using the Umaes' own tactics against them. Feigned retreats leading to encirclement *by* the Hybrids often led to enormous losses for the Umae. Initially, Dom was enraged by such defeats. But then he came to learn that no Hek corpses were being found on the battlefield anymore. Dom reveled in the satisfaction of that victory for ellipses despite the increased challenges posed by the Hybrids. It was simply a matter of time now for him as it had always been. Dom was confident that his long lifespan would enable him to see the entire planet cleansed of Hybrids.

Joba raised his hand and brought more troops forward. They were a long way away from High Hill. He had already sent a messenger back to Dom advising him of the imminent battle. Dom sat in the midst of his armies like a hungry spider, sending his troops out in radii from his base in search of the Hybrids.

Joba's soldiers readied their shields. These were the children of Dom's wisdom grown into men. Their bodies were strung with muscle like woven rope. They could not be denied the satisfaction of battle for too long lest they seek release in wanton assault and murder against their own people. Dom had to keep them in the field where they wanted to be. With such strength and temperament, they were nearly as much a threat to their own settlement as they were the Hybrids. They had besieged the Hybrids for forty-six rotations. The Umae had entirely encircled the Hybrids' hilltop fortress allowing nothing in or out. By now they would be half-starved and desperate for water. Weakened by want, the Hybrids would be helpless to resist. The small advantages of their elevation, and their paltry wooden wall, would be minimized. Turning towards his soldiers, Joba pumped his arms up and down and incited his men into an inflamed fury. Soon, one less threat to the Umaes' survival would be eliminated. Dom would be pleased.

The Hybrids would attempt to slow them down with spears and javelins. They might also roll rocks down the hill. Joba had seen all of this before. The Hybrids were pitiful. Never had they ever sought battle on equal terms. Their treachery was matched only by their cowardice. They did not deserve death. With a wave of his hand, Joba sent a wave of howling Umae towards the palisade. They had seen no signs of the Hybrids for several rotations.

The Umae focused on a huge gate in the center of the wall facing them. The wall itself was easily twice as tall as they were. They reached the base of the wall without any signs of resistance. The mass of Umae eagerly stalked the perimeter of the Hybrids' structure desperately seeking a means to get to their enemy. Another group of Umae had finished dragging a large tree trunk up the hill. A third group replaced them and lifted the tree trunk off the ground. Aiming its end at the gate they reared back and drove it forward. The sound of the collision between the tree trunk and the gate reverberated back down the hill. The men reared back and drove the trunk into the gate a second time. The gate gave slightly and the Umae could hear the thrilling sound of the wooden gate splintering. They redoubled their efforts. After smashing the trunk into the gate six more times the gate finally gave way. The Umae shoved the tree trunk to one side and began pushing the shattered remnants of the gate inwards to create more space for their intrusion.

On the far side of the gate was an enormous hallway formed by two walls that ran parallel to the invading horde. The floor was nearly covered with Hybrid corpses. Their stench swirled in the nostrils of the Joba's warriors as the buzz of flies filled the hall. Joba raised his hand and bid his men to be silent. While their fervor made this almost impossible, they restrained themselves enough to be able to hear the sound of a Hybrid horn sounding from somewhere beyond the far wall. Joba responded by reforming his troops into lines five wide, as many as the passage would allow. The front ranks wielded short iron blades while the others readied long spears. Once the formation was complete, they strode forward.

This was by far the largest Hybrid fort Joba had ever seen. That simply meant that the siege was more likely to be successful. More people needed more food and water. He and his men marched down the hallway towards a set of double doors. The front rank shoved the doors open and the Umae poured through like a column of ravenous ants. There was a huge square room on the other side. The far wall was lined with two dozen Hybrids sitting closely packed together. Many had placed their head on the shoulder of the fellow next to him. They didn't move. In their center stood a single man clinging to a long spear adorned with feathers. Bright paints colored his face and arms. The room itself appeared to be a storage area of some sort. There were odd boxes and bags strewn about. Amongst the ceiling beams hung a variety of clay jars.

Joba smiled. "Evil fools," he called. "Judgment is upon you! Upon you all! Our final victory is near at hand."

The Umae in the back ranks hooted with glee as they began to spread out behind Joba and the others. "Perhaps," replied the Hybrid. His response was in Umae.

"You befoul our tongue!" spat Joba. "For that, you die last and will watch all of your kind die before you."

"My kind aren't here," said the man. "At least not very many of them." He covered his mouth to cough leaving his hand stained with blood.

Joba regarded the figures lined up against the wall. "Dead already," he observed.

"Oh no," said the man. He glanced down at the figures along the wall. "But their deaths will serve our people well." The men seated against the wall slowly rose. Each effort was plainly a struggle as most of them appeared to be gaunt and weakened.

Joba cackled. "A pity, so weakened by a lack of food. And of water. Not that they would have posed much danger to us anyway!" The Umae laughed.

"They will be your death," said the man. Again, he began coughing up blood into his hand. "Weakened yes, but not for a want of food or water. A few of our number brought a special weapon with them to this place. They shared it with us as we will now share it with you."

Joba scowled as his men continued to laugh and throw insults at the Hybrids. "What weapon? I'm sorry, but are you mad?"

"A great weapon," croaked the man. Joba thought he could see a wry grin on his face. "Oh yes." The man placed his hand against the wall and found a length of rope, pulling it downwards before Joba and his men could react. The clay jars overhead all tipped sideways in near synchronicity, showering a vile green-brown liquid on the unsuspecting Umae below.

"Aggggghhh!" cried one as he tried to wipe the grotesque substance from his face. "It's foul!"

Other Umae in the room scrambled to rid themselves of the odious liquid. The stench in the room was becoming overpowering and many of the Umae began to wretch.

"Kill us now," said the Hybrid leader as the other Hybrids sat back down. "But as I said before, we are already dead. Today is our victory."

Joba screamed. His Umae lost all discipline and charged at the Hybrids, hacking into them with their short blades and puncturing them with their spears. Their leader smiled as he fell.

The remaining areas within the Hybrid fort were empty. There were no bodies, no weapons, and very little food or water remaining. Joba did his best to conceal his confusion from the other officers present.

"The cowards have fled," he announced. "And none too soon for them."

"But how?" asked another. Binzel was a product of Dom's wisdom. He was a thickly built, powerful Umae who longed for the wanton slaughter of combat. "They were surrounded."

Joba paused. "Their scouts must be more effective than we gave them credit for," he replied. "That is the only explanation as to how they got everyone out of here before we could begin our siege. But we are victorious anyway. They cannot win through clever retreats. We will continue on, find more of them, and slaughter them as well."

Binzel looked down at his clothing. "We must return to the creek," he growled. "Whatever this is that they dumped upon us is disgusting. I'll bear it no longer!" The soldiers listening to their captains rumbled in agreement.

Joba nodded. "Indeed," he said, noting the gooey green mess on his own clothing. "And Dom is due a report. We should return to High Hill. There we can tell him the news of our victory and perhaps receive one of his personal rewards." A leering grin arose on Joba's face as he imagined his sleeping chambers stocked with beautiful young Umae girls.

Binzel grunted. "The rotation is yet young. Let us challenge the Sky Fire. The sooner we depart, the sooner we can be rid of this awful stench. And the sooner we will receive our rewards!"

The other men laughed with approval.

Joba took Binzel by the elbow and led him a short distance away. "You have earned the ears of the men," he said in a low voice. "I have been doing this for many, many ellipses. Dom may deem me deserving of retirement. If he does, I suspect you will be placed in command."

Binzel pulled his arm away from Joba. "Deserving? No, not in the way you might think."

Joba didn't see the short blade before it entered his guts. He drew a short, quick breath as he stumbled forward, trying to hold himself up by seizing Binzel's shoulders. With a firm shove Binzel sent the older man to the ground, delighting in the sight of Joba's blood pouring out onto the stone. Binzel placed his foot on Joba's shoulder and rolled him over onto his back.

"Co-ward," gasped Joba.

"Fool," replied Binzel. The other men had barely reacted at all. Binzel took the cuff off Joba's sleeve and pulled it up towards his elbow, exposing his forearm. It was covered with large, irregular brown patches. "A weak fool at that," he noted. "You cannot bear the strength of the Sky Fire. Those who do not bear the wisdom of Dom fall victim to it eventually. You are sickly. We will purge you and your kind from our tribe. We will then rule the day as well as the night. The time of the Umae has passed. The progeny of Dom's wisdom shall take your place."

Joba was unable to focus as the loss of blood was making his head spin. "You....will regret.... I have served Dom well."

"You have outlived your usefulness," said Binzel plainly. "Consider that as you wait for death."

Binzel stepped over him and walked to the far end of the chamber. The other men followed along behind him with barely any regard for Joba at all. They filed back down a long hallway and out into the glare of the Sky Fire.

"The creek?" asked one.

"The creek," echoed Binzel. "And after we have cleaned ourselves, we will celebrate our victory."

Binzel's army retreated back down the slope leading to the palisade. There was a small stream just a half rotation's away. That would be their destination before returning to High Hill and advising the great Chikwa of their victory.

CHAPTER 19

Lin's nostrils twitched in protest. The lining of her nose stung as the stimulant filled her repose tube. She blinked hard and saw the display activate in front of her.

'Fifty-eight percent,' read the panel. She tried to focus. Her arms and legs tingled as the repose tube stimulated her extremities with a mild electrical current.

"Status?" she asked.

'Seventy-one percent,' read the panel.

She wiggled her fingers and toes. The air inside her tube smelled cleaner now as the concentration of stimulant was reduced.

"Are we home yet?" she asked.

'Ninety percent.'

Lin closed her eyes and waited. She knew it would be just moments before the repose chamber opened. Now that her thoughts were clear, she remembered the studies she had assigned to sim-Jack. She was anxious to check the results.

'Heart rate accelerating at an abnormal pace', read the panel. 'How do you feel?'

"I'm just eager to get out of here," said Lin. "I have some important work to check on."

The panel went blank for a distressingly long time before the chamber finally slid open. Lin stepped out onto the deck of the ***Aurora***, gently shaking the tingling sensations from her legs. She squatted halfway to the ground to check her balance. Confident she was fully recovered, she moved to the other repose chambers.

Eve was still in the chamber next to hers. Her chamber gave no indications of repose termination. Lin checked the chronometer on the outside of the chamber. Eve wasn't due to awaken for another twelve rotations. Now fully awakened, Lin recalled her diagnostic plan for Eve. Eve needed as much repose time as possible to fully recover from her interface with the satellite. Lin didn't believe that she would need Eve to check the results of the RTS studies, so she had opted to keep Eve in repose a bit longer. If the need arose, Lin could always awaken her manually.

Frell was still in his tube but the one next to him was missing entirely.

"Sol?" whimpered Lin. The aged Life Agent had been in the tube between Frell and Eve. Now that tube was gone leaving only a vacant tube slot in its place. Lin had no idea which tube Jack had used. For all she knew he had a separate tube in his private chamber.

She went to Jack's door and began beating on it. "Jack! Jack! Come out here!" She pounded on the door until her hand smarted. "Now!"

She stood back and glared at the door, trying to will it open. She heard a brief clinking sound before it swung outwards towards her.

Lin gasped. The man who stepped out from behind the door was enormous. Her eyes were level with the base of his sternum and he was easily twice as massive as she was. His hair hung past his shoulders and showed signs of gray streaks.

"What do you want?" snapped Jack. And it was Jack. The sound of his voice, albeit a bit deeper than she remembered, left no doubt.

"You.....what did you do?" asked Lin. "You're........"

"Better," said Jack. "Bigger, stronger, better than ever," he said happily. "All part of my plan."

"But how? How did you do this?" asked Lin as she tried to take the sight of him in, attempting to understand his transformation.

Jack merely shook his head. "You aren't the only one with Life Agent expertise," he said. "My expertise is almost unlimited. You hardly helped this at all."

"What do you mean?"

Jack pointed towards a nearby panel. "Your test results await. I'm confident I was as helpful as you were hoping."

Lin shoved her disbelief aside and rushed to the panel. "Sim-Jack, summarize the test results."

"No, no, no," said Jack. "It'll do no such thing. How about real Jack? I'll do it for you. Complete success. The compounds you tested are the key to manufacturing a cure for RTS. My sim managed the contingencies perfectly which isn't really a surprise. I'll give you a minimal amount of credit for setting up the studies themselves." He glanced over at Eve's response tube. "She will be thrilled. She will get to see Min again. And Min will get to see......this!" he added flamboyantly as he spread his arms wide. "The *Aurora*! The greatest Journeycraft in the history of the Umae and it isn't even close! It was the greatest when we left Haven, but I've spent the last ten ellipses expanding its capacities even more. I can't wait to see the look on his face when he sees it!"

Madness. Lin looked into his eyes and realized that any remnant of sanity he had clung to before was gone. Ten ellipses. He had been alone, basking in his own superiority, for ten ellipses. The last vestiges of the old Jack were gone. In his place stood a mad, raving giant.

The empty repose tube slot jerked her back. "Where's Sol?" she asked quietly. A tendril of dread began to slowly weave its way upwards from the pit of her stomach into the back of her throat.

"Didn't make it," said Jack flatly. "I was going to use him as a test subject, just like you said, but he didn't last the trip. But Min's life signature is still transmitting, so we still have him. All this theory isn't much good without a practical success, right?"

"No!" Lin screamed as she rushed at Jack unsure about what she would do once she got to him. She brought her hand back to strike at him but found her arms pinned to her sides. His speed and strength belied the apparent number of physical ellipses his body had experienced. He held her at arms' length and bared his teeth at her.

"Never. Again," he snarled. His hands felt like they might crush her bones.

"Let me....go." She tried twisting away but that only amplified the pain.

Jack shoved her backwards. "I fail to understand your reaction. You told me yourself your intent was to use Sol as a test subject. I attempted to cure him ten ellipses ago. Sim-Jack agreed with my plan. After all, the sooner we brought him out, the better."

"You used a Tech Agent sim to make that decision!" barked Lin. "Why didn't you wake me up! How did you even know what to do?"

Jack shook his head. "Arrogant woman. You think you are so superior. He was gone before we even took him out of the tube. His death is your fault, not mine."

Lin turned away from him, her body shaking with sobs. It was true. Sol's chances of surviving this last trip to Uma weren't good. But she hadn't had any other options. Had she?

"Get away from me," she said quietly. "Just go."

Jack sneered. "I was planning on it. I have work to complete. On a project that will be entirely successful. We will be at the Citadel in sixteen rotations. I would suggest that you familiarize yourself with the use of my new medication. It would be awful if you killed Min, too."

She kept her back towards him as she fought for control. She stared at Eve and Frell in their repose chambers.

"No one else is going to die," she said firmly. "You can count on that."

CHAPTER 20

The sound of the rushing waters was a great relief to the soldiers. While they had spare clothes and had disposed of the ones soaked by the Hybrid attack, the foul mixture seemed to cling to their skin and hair. The stench roiled their stomachs although none were willing to admit that weakness by complaining about it. They quietly filed into the creek, stripping off their clothes as they went. Then they went to work to scrub themselves clean. When they had finished, they stood naked beneath the Sky Fire to dry off. None of them bore the sickly brown splotches that had been present on Joba's arms.

"Come!" cried Binzel. "We will bear great news to Chikwa. And we shall be regaled for it!" The men cheered and pumped their arms, slapping one another on the back. Some had benefited from Dom's generosity before and had spoken frequently of the experience. Those who had not burned with anticipation.

Their eagerness drove them across the prairie towards High Hill. It was a three-rotation march that the men completed with vigor. Dom's scouts met them near the edge of the settlement and asked them for news.

"Victory!" proclaimed Binzel. "Complete and total victory. Share the news with Chikwa. He will be pleased." The scouts bowed quickly and headed off towards the mesa. The men continued their march, but now at a more leisurely pace. They wanted to allow Dom plenty of time to prepare for their arrival. As they entered the settlement area the scouts reappeared.

"Dom is indeed pleased!" they proclaimed nearly in unison. Another group moved up behind them bearing a number of wooden boxes. "He has decreed that all of you will be gifted with a deerskin tail. The greatest reward of all!"

The deerskin tail. It symbolized that one had participated in a great victory. It merited a great reward for the bearer. The soldier bearing a deerskin tail had his choice of any woman in the settlement. The wisdom of Dom was perpetuated with all of its virtues. The men gathered around while the scouts handed out these great spoils of victory.

"Ha!" cried Binzel. "Go men! My brothers! Go and claim your just reward!" The men cried joyfully back to him as they began to disperse, each of them with a particular woman or girl he had been yearning for on his mind. So much waiting would come to an end.

"Chikwa would speak with you, Binzel," said one of the scouts.

Binzel frowned. "Why? Am I not to be rewarded as well?"

"Yes, of course," said the scout as he handed Binzel his trophy. "But he insists that you delay your revelry until after you have met with him."

The Umae captain growled. "Very well," he grumbled as he snatched the skin from the scout's hand. The warrior strode off in the direction of the Chamber of Waters.

Several times his attention was drawn by one of his men claiming his prize. Dom's decree was known and it was observed by everyone. Joining with a victorious warrior was not the worst thing that could happen to an Umae female. Even those with mates understood the importance of these military victories. The Hybrids must be defeated. The women did not fight but contributed as they could, nonetheless.

The guards outside the cave barred Binzel's entrance until the Sky Fire dropped below the horizon. Dom would see him on Dom's time. Finally, Binzel was permitted entry to the cave and found Dom seated on his throne.

"Total victory," said Dom firmly. "Tell me what you mean by 'total'".

Binzel swallowed hard as he genuflected at Dom's knee. "All dead. All of them. Joba included."

Dom smiled. "How many?"

The memories of the sparsely manned fortress flooded Binzel's mind. He considered lying, but the risk......

"Fifty," he said finally.

A shadow rolled across Dom's face as he stepped down from his throne. "Fifty? Fifty? Was there not a fortress there?"

"Yes. And we encircled it as soon as we arrived," said Binzel defensively. "There would have been no escape."

"They knew you were coming......" growled Dom. His face flushed. "So tell me, why did they leave fifty inside the fortress? Hardly enough to provide any resistance." Dom was prowling around his throne room like a caged wolf.

"They could not escape," said Binzel boldly. "They were suffering from our siege."

"How long was the siege?"

"Forty-six rotations."

"And was there food and water inside the fortress when you entered?"

"Some."

Dom turned slowly towards his lieutenant. "What makes you believe they were suffering?"

Binzel tried to moisten his lips. "They were coughing. And bleeding from their noses and mouths. They could barely stand and were not able to fight."

Dom trembled slightly. "So they just stood there and let you kill them?"

"Well," said Binzel with a chuckle. "No. They tried to pour this foul matter on us from buckets hung from the ceiling. The stench was as death. It was truly pathetic."

Dom stood up straight causing every bone in his back to pop. The sounds echoed through the chamber. "And these men, your soldiers, these victors, they now revel all throughout MY settlement?"

"Yes, my Lord. As I plan on doing as well." The sound of Dom's fingers sinking into Binzel's sternum were dwarfed by the wail of agony released by the soldier. He futilely pulled at Dom's arm, trying to free himself. "I don't........" The air was quickly racing from his lungs.

"Fifty?" began Dom. "FIFTY?" He slung Binzel across the room, slamming his body against the far wall. Binzel slid to the floor and remained motionless. "You killed fifty. They have killed THOUSANDS!" He raised his arms in rage and released a bellow that shook the walls and rang down the slope, causing his followers to scramble for cover.

The club's head crashed into the child's skull, spraying blood and gray matter against the exterior wall of the dwelling behind it.

"All of them," snarled Dom. "Every last one. You will be vicious or you will be next."

The club-bearing Umae nodded and began looking around for other victims. Fear-drenched cries intermittently rose and were abruptly cut off from all around the settlement. Some of the Umae victims attempted to run, but most were suffering from the effects of the plague. It robbed them of their strength rendering them easy targets for Dom and his progeny.

One woman ran to Dom and, sliding to her knees, clutched at his legs.

"Chikwa, why? Why is this done?" Her expression was of confusion twisted by despair.

Dom seized a handful of her hair and swiftly drove his knee into her face. The front of her skull collapsed leaving only a gaping pit of bone and blood. He let her body flop limply to the ground.

"My progeny, to me! Now!"

His voice echoed throughout the settlement, easily overwhelming the chaotic sounds of pursuit and death. He stood and waited as his warriors reached his position. Their garments were all spattered with blood. Their clubs and spears bore the stains of mass murder. Aside from an occasional mewl of suffering, the settlement was quiet.

The Umae warriors knelt before Dom in ten rows of ten. There were no women in this group. The slope leading to Dom's cave, and the spring, rose up in the distance behind him. Dom waited before speaking, basking in the near palpable fear radiating from his soldiers.

"It seems I was mistaken," he said plainly. "Despite your lineage, you aren't nearly as smart as I'd hope you'd be. The Hybrids not only bested you, but they did it without you even recognizing the scale of your defeat. And now........."

He spread his arms out, beckoning at his settlement. Bodies were strewn about in every direction. Many still had blood dripping from their noses and mouths. Their clothes were stained with a green, slimy substance. A few Umae could be seen near the horizon struggling to escape.

"Now everything I built here is gone. Because of you. But........," he continued as he began to stroll back and forth, "what am I to do?" The full moon was just over the horizon. "I do not have a suitable mate with which to reproduce. And this is the fault of the Umae. I thought my seed would be adequate to remedy the multiple shortcomings of your species, but I was wrong." He seized the nearest soldier by the shirt, causing the man to nearly cry out. "You are far stupider than I'd ever imagined, and all of you are the products of my 'wisdom'".

The memory of the Umae women, those who bore all of his offspring, momentarily softened his expression.

"But no matter. I still intend to have my revenge. And I will have it by claiming the Umae line as my own." He released the soldier and took a step back. "From this rotation forward, any man who is not of my line shall die. Umae, Hybrid, it makes no difference. And any woman we find shall bear my seed. Umae, Hybrid, all the same. And that offspring shall bear more offspring, and that generation more generations, until every Umae and Hybrid that remains on this planet shall be of my line. And that will be my victory."

The men continued to kneel, each with questions restrained by fear.

Finally, one spoke. "Sire, what has happened to these people in this settlement? They were dying yet we were not."

Dom considered the speaker. "Rise. What is your name?"

The soldier slowly stood up. "Masor," replied the man cautiously.

"Masor," repeated Dom. "The brighter ones of you surely had questions about what I've said, although I'm confident that number is small. Of those, only you had the courage to speak. So you are relatively bright and relatively brave. You shall be my lieutenant."

Masor nodded almost imperceptibly. "I will serve you well, Sire."

"Or your service shall be brief," noted Dom. "But your question. 'What happened?' Here is what happened. The Hybrids tricked you all. They made you bring their plague back to my people. What they couldn't accomplish by feat of arms they did with illness. An admirable tactic, but one that will cause their extinction. It was only my blood in your bodies that saved you from that grim death. But I had to be certain that was the case before acting to remove the.....less fit....from our group. From this point forward, only my line will survive. We will not enslave the others, we will slaughter them. And any female we find will become a cow for my progeny. The men die and the women breed. And when the women are done breeding, they die as well. And as my line grows, our army grows. And as our army grows, we become invincible. And my vengeance against the Umae and their pathetic Journeyers will be complete. For one day, they will return with a plan to save their race. And when they do, they will discover that their race has become my race."

Masor rose and turned towards the other men. "Guide us, Sire! Your will is ours!"

Dom looked out over the darkening plain at the distant figures attempting to drag themselves to safety.

"Kill them. Kill them all. Every last member of this settlement shall die before we see the Sky Fire again. And then, my children, my wrath will spread like fire through a kindling forest. Death, misery, defeat. All of these shall be brought to the Umae and the Hybrids. And then I will wait until my rotation of reckoning when the Journeyers return. I will show them what I have done, what they have caused, and then they will experience the most torturous deaths I can imagine. And," he added as a deep ebon swept across his countenance, "I still have a lot of time to imagine."

Dom and his army stood atop a bluff overlooking an Umae settlement. They had marched long and hard for innumerable rotations, leaving High Hill far behind. The rudimentary structures spread out below them had no protective walls whatsoever. As the moon cleared the horizon, Dom could see no signs of even a posted guard.

"Either this is the stupidest group of Umae I've ever encountered or my reputation has not spread as far as I thought," he noted. "No walls, no guards." His cohorts were not able to see in the darkness like he was. Dom made a mental note of the shortcomings inherent in creatures with impure blood. The face of the moon taunted him as it stared downwards. Indirectly it was the reason for this genetic failure. The physical pain he had endured upon the loss of his mate echoed in his chest.

"What is it My Lord?" asked Masor. "You appear aggrieved."

"'Aggrieved'?" growled Dom. "Oh I'm aggrieved all right, in ways you can never imagine. But I'm a generous conqueror. My aggravation will be shared."

Turning around he saw his men lined up in twenty rows of ten awaiting his instructions. Although there were subtle differences, they all looked essentially alike. Taller than an average Umae and more muscular. And they all shared a zeal for slaughter. Yes, Dom had had a positive impact on the genetics of this species. And it was about to get even more pronounced.

"What is your will, Sire?"

Dom pivoted back towards the settlement. "Simple. Kill the men and children and seed as many of the women as possible. But don't kill the women. And don't take them captive. These Umae don't know it yet, but we will be doing them a favor."

"But Sire, we are not going to subjugate them?"

Dom broke out in a broad smile. "Oh yes. I will subjugate their bodies. I will subjugate their entire species. Their women will bear my progeny. And that progeny will be like you. They will follow their best instincts, ensuring that my line of Umae will replace theirs. And then, when they come back......."

He looked up at the moon and glowered.

"Guide our way, Lord."

Guide our way.

Dom broke into laughter. His soldiers glanced at each other curious as to what had triggered Dom's outburst. None were brave enough to ask. Once Dom regained his composure, he raised a single clenched fist causing his warriors to snap to attention.

"Now," he spat. "Be relentless. Be merciless. But most importantly, be fertile. All men and children die. All women become my vessels, either directly or not." He lowered his fist. "Now!"

His men circled around the bluff finding a path to the bottom. Dom led them forward in the darkness, guided by the bright face of the moon. He and his soldiers swept into the nearest structures, easily tearing them apart. The Umae inside raised surprised cries of terror as they attempted to assess this threat. Other Umae from more distant dwellings emerged only to be overrun by the warriors who had raced past the perimeter dwellings. Although some of these Umae were armed, they were no match for the speed and tenacity of

Dom and his minions. The men stood and died while the women attempted to flee with babies in arms.

Their flight was hopeless. Even given the head start offered by the slaughter of their mates, the women were quickly caught and encircled by Dom's horde. They huddled in the center of the circle clutching their children. A few young Umae boys possessing perhaps ten ellipses attempted to stand defiantly with their spears in between their mothers and their attackers. The tears on their cheeks and the tremble of their hands betrayed the authenticity of their valor.

"Hold!" called Dom. He stepped towards the boys who instinctively raised the points of their weapons. "You seek to defy me?" he asked plainly.

The boys were breathing hard with many of them openly sobbing.

"Yes!" said one. He was smaller than the others but nonetheless took a couple of steps forward towards Dom. "You are a child! You are no better than we are without all of these men!"

"You are wrong in so many ways," said Dom calmly. "There was a time when I would have found pleasure in correcting your misunderstandings, but that time has long passed. Tell me, is one of these women your mother?"

The boy glanced behind him, betraying her identity.

"Fendis, come here," said the woman. Fendis held his position.

"I'll tell you what, Fendis," said Dom. "You have shown courage today. In honor of your memory, I will seed your mother myself."

Fendis bit his lip and redoubled the grip on his spear. He wasn't entirely certain what Dom had meant, but those words brought a tightness to his chest. "You must kill me fir-----."

Fendis did not get the opportunity to finish his challenge. Dom had driven the butt of the boy's spear backwards staking Fendis through the sternum. The child collapsed as a rush of foaming blood poured from his mouth.

Dom's warriors closed and made swift work of the children, ripping them from the arms of their screaming mothers. The women were herded roughly into a smaller circle of warriors. Again, Dom emerged at the front of his army. "You there!" he said pointing to Fendis's mother. "Come here."

The woman sneered and spat on the ground in his direction. Her hands and clothing were stained with the fresh blood of her child. "Never!"

"Oh I think you will," he said in a low, slow voice. His words slid over her thoughts like a warm sea tide. "Come to me." She stood up somewhat uneasily on her feet and clambered forward. As she reached Dom, he took her by the hands. "Now. Take them! All of them!"

With a roar Dom's men fell upon the Umae mothers. And the great march of his legacy took its first steps into the future.

Dom and his band swept across the land like a raging pestilence, overwhelming Hybrid settlements unprepared for the speed and viciousness of the tactics they employed. The pattern remained the same. Dom would scout ahead and locate a suitable gathering of Hybrids. He would then devise a plan intended to slaughter all of the males in the population, including children, so he and his army could violate the remaining women before letting them flee into the wilderness. Dom's reconnaissance was so stealthy that he was able to remove any scouts or sentries employed by the Hybrids allowing his warriors to strike without warning.

His Umae worshipped him. He provided them temporary relief from the bloodlust that flowed in their veins. He provided them opportunities for repeated carnal assaults against powerless victims. And the more Dom provided them, the more they revered him. And Dom, in turn, basked in their adoration. But each time the moon began to wax in the nighttime sky, and its pale face mocked him from above, he became more and more incensed by the reality that his victory could never be anything more than incomplete.

"Stop!" commanded Dom. "We stop here."

As always, his men marched in neat ranks behind him. They stopped but looked at one another questioningly.

"But my Lord," said Masor, "the rotation is still new. The Sky Fire is fresh. Do we not have time to display our prowess?"

A chill shook the lieutenant's spine as Dom turned and looked up at him. "We stop here. Do not question me again."

Masor lowered his head. In a quaking voice he replied, "As you wish my Lord." He turned towards the rest of the men. "We stop here!" he called. "Refresh yourselves. Scouts! Seek out fresh water. We shall remain until our Lord guides us forward."

The water scouts began to break formation as Dom also turned and raised his hand. "No! You will all stay here! Battle has found us!"

A stir of confused commotion rolled through his ranks. The men casually studied the horizon but saw nothing. They gripped their spears and clubs and stood quietly.

A slow smile grew on Dom's lips as he saw the rows of men standing before him. Somewhere beyond the farthest rear rank came a clamor followed by the sounds of clashing arms.

"Behind us!" cried Masor. "Turn now! Turn and attack!"

As Dom's warriors turned to assess their attackers, they realized that a wave of Umae warriors had overwhelmed the far end of their column effectively crossing its "T" and enveloping it on both sides.

"My Lord?" asked Masor. "Your command?"

"Fight," spat Dom. He remained still and provided no indication that he intended to join the fray. "As if your life depends on it."

Masor's eyes widened momentarily before he hefted his spear and sprinted down the line of warriors. "Turn the column!" he cried. "We must turn........" A sling stone split through the tumult of battle, cleaving his forehead wide open. A spray of blood and bone erupted from his skull before he toppled backwards to the ground. Dom paid his corpse no regard as he stood watching the developing battle.

The attackers had surrounded the end of the column on three sides. Dom's men tried frantically to swing the other end of the column around to attempt a counter-envelopment. The Umae attackers were deadly accurate with their staff slings and many of Dom's warriors fell before they could complete their maneuvers. Dom's men looked back over their shoulders at their Lord, confused by his impassive disinterest.

Finally one of his warriors raised his club in the air and released a war cry that rallied his comrades. Dom's men fought their way through to their surrounded comrades with many on both sides falling. But the numbers of the enemy were too great. Dom's men had lost too many in both the initial assault and in the volleys of sling stones that had rained down on them. They could not hold their line as the weight of the enemy bore down on them.

"Lord! Lord save us!" they cried. But Dom remained unmoved. They were steadily cut down by the attackers until they attempted a final failed attempt at retreat.

Seeing his men flee, Dom surged forward.

"Cowards! Hek-blooded curs!" he screamed. With incredible speed he swept through those in retreat, breaking necks and shattering skulls with his bare fists. None escaped. Every last member of his army had fallen. Now it was only Dom and a steadily advancing group of Umae attackers. He stood calmly as the victors marched towards him.

"My Lord," said one as they approached. All of the Umae warriors knelt down and bowed their heads.

"The strongest," Dom noted casually. "You will not rely on my might. You will be strong of your own right. Unlike these weak mounds of carrion before me. I will grow our numbers like weeds and we shall spread like the sea from a broken dam."

Dom chuckled to himself as he imagined that image.

"But some of you will ravage this land without me," he continued. "You know what is to be done. It is in your blood." His eyes narrowed. "I am in your blood. And my blood shall be all that remains on this accursed planet. The people of Uma will know no peace. They will know nothing except what my blood teaches them. War. Death. Suffering. And they will know it forever."

He turned his attention to the horizon. The Sky Fire was high overhead now. But by the end of the rotation, the moon and its torturous face would appear and Dom would welcome it. He wanted to tell it about the suffering to come and about how the face in the sky would be powerless to do anything about it except look down for eternity and weep

CHAPTER 21

Lin didn't wait for Eve's repose to end before she began the manufacturing process. The first batch of medication was ready before the Data Agent emerged from her chamber. As Lin had expected, the long repose cycle had done Eve a great deal of good. Her nervous system had been given a lengthy period of low activity and now was back to normal. From her examination of Eve, Lin couldn't detect any signs that Eve had ever incurred any neural trauma at all.

Lin's impatience was justified. Eve culled through the data generated by the studies and confirmed that the newly manufactured medication would, in fact, cure RTS. Lin struggled to share the news about Sol. Eve was so upbeat about her reunion with Min. It seemed cruel to tell her about her dear friend's passing.

"I'm going to wait as long as we can to awaken Sol," Lin explained. She was such a horrible liar. She felt confident that Eve suspected something was amiss. "I want to make sure the timing and the dosage of the medication are ideal. He has a lot of ellipses and the RTS has taken a toll on him. I was wondering if he'd survive the trip at all. Jack suggested that we transfer him to his private chamber. He said the tube docks there were more efficient."

She had managed to choke out that last phrase without losing her composure. Perhaps Eve's emotional distraction was preventing her from seeing through Lin's ruse. If she suspected anything, she wasn't saying so. Lin accepted that set of circumstances because it didn't require her to repeat her lie over and over.

Shortly after Eve awakened, the *Aurora* notified everyone that Uma would be available for viewing on the ship's Alpha display very shortly. Jack emerged from his chamber for the first time since Eve's repose had terminated. She looked him over briefly but didn't seem to be too disturbed by his massive size. In fact, Eve wasn't concerned about much of anything until the planet came into view.

"Oh my sweet moons........" said Lin in a hushed voice. Uma loomed on the display, its former dark gray cloud cover now replaced by swirls of light white and azure. Eve began babbling and clutching at the sides of her head.

"H-hurts......" she muttered. "Uma......."

Lin rushed to her side and immediately donned a pair of examination circlets. "Just hold still, hold still," she implored. She slid the membranes over her hands and held them to Eve's head.

"What's her problem?" asked Jack coolly.

"Shush!" Lin closed her eyes as she sorted through the signals the membranes were feeding into her nervous system. "Quickly! My pack!" she said to Jack. "I need it."

Jack glanced around and saw a med pack near another panel. He retrieved it and set it down next to Lin. Lin removed a small tube from the pack and squirted a pasty substance from one of its ends onto her hand. She then began massaging the paste into the back of Eve's neck.

"Easy. Just relax," she said in a quiet voice. "You are fine."

Eve had been rocking back and forth, her face twisted in terror. As Lin went to work, Eve slowly began to calm down. Soon, she was sitting still, leaning against Lin.

"I've seen this," she said finally.

"I know," said Lin. "You are finally having a flashback. The sight of Uma must have triggered it. Can you talk about it?"

Eve nodded. "Energy wave. It washed over the entire planet. It came from the satellites. The atmosphere was radically altered. It's done. We are too late."

"No, we aren't," said Jack. "I've had the **Aurora** doing readings on Uma since it came into range. The sensory range on my ship is much longer than its visualization range. The atmosphere has been altered, but not so much that it can't support the Umae. If the aliens attempted to change the atmosphere as a part of their breeding strategy, they failed. Tell me, Eve, what else can you remember?"

Eve closed her eyes and tried to focus. "The satellites moved, forming an arc about the planet. They generated a wave pulse that moved over the planet's surface as it rotated. But, there's something else?"

"What, Eve?!?" demanded Jack. "Tell me!"

Lin scowled at Jack. "Be patient! Her subconscious suppressed all of this. This is extremely difficult for her."

Jack stood over Eve, rocking back and forth.

"A bubble. A bubble of energy. It was between the satellites and the planet. The wave had to go through it. No. Two bubbles."

Jack pondered this revelation. "The bubble changed the wave," he said finally. "But where did the bubbles come from? The aliens wouldn't have sabotaged their own plan. There is only one possibility. Min did it."

Eve's eyes brightened at the mention of her love.

"So you are saying that he did something to save Uma?" asked Lin.

"It's the only logical conclusion," said Jack. "The aliens attempted to modify

Uma's atmosphere like they did on Haven, and Min did something to at least mitigate the effects. I should be able to determine what happened once we reach the Citadel."

Lin squeezed Eve slightly before glancing back up at the display. As it panned away from Uma, the full moon came into view. "What…..?" Lin was out of meaningful questions.

Jack completely lost all composure and burst into a violent fit of laughter. "That's absolutely fantastic!" he declared once he could catch his breath. He tilted his head to better observe the crater pattern on the moon's surface. "Min drew a face on the moon! A face! He stopped the aliens and then he drew a face to taunt them! It's glorious!"

"That doesn't seem like something he would do," said Eve soberly. "Not at all."

"Well he did," continued Jack. "There can be no doubt. I guess he's not so pitiable after all. Knowing me must have given him some extra juice."

"What else do your scans say about Uma?" asked Lin.

Jack turned and started walking back towards his chamber. "No time for that now! All part of the plan. Ship, set course for the Citadel!"

"Touchdown coordinates adjustment required," said Jack's disembodied voice. Jack had rewritten the *Aurora*'s base operational directives so that its Alpha panel's audial communications function used his voice. Lin had only interacted with sim-Jack during the trip to Haven and now felt fortunate that she hadn't had any sort of access to the *Aurora*'s command functions. She couldn't imagine hearing Jack's voice constantly for all those ellipses she was out of repose during the journey. "Debris detected in the proposed landing zone."

"What sort of debris?" asked Jack.

"It consists of two major elements. The first is metal fragments consistent with a heavily damaged shuttle craft. The second is an array or inert mech units spread about the surface of the moon but oriented generally in the area nearest the Citadel."

Jack smiled. "That's how he made the face," he said confidently. "He used the mechs. But why? I can't wait to ask him. I assume you haven't been able to affect access to the Citadel's docking bays?"

"Negative. No response from the docking bays."

"Then just set me down as close as you can get without risking any contact with the debris. I don't have time for repairs now."

"Adjusting landing coordinates."

211

Lin got up and began walking towards a large storage locker. "This will be a first for me," she noted. "I've never had to use a vacuum suit."

"I have," said Eve. "Lots of times. I'll be happy to help you anyway I can."

"Well I haven't used one, either," said Jack. "But I won't be needing any help. I've redesigned the *Aurora*'s vacuum suits. They are foolproof. Even you two should be able to manage." He stepped past Lin, brushing her aside, and withdrew a suit from the locker. Even his vacuum suit was enormous. Lin glared at his back before removing two more suits and handing one to Eve. In no time, the three had slid into them and were ready to depart.

A motionless mech sat steps away from the bottom of the egress ramp, a shovel full of moon dirt held high over its body. All around the stretch of plain between them and the Citadel they could see a number of identical machines all as still as the moon's surface. Jack knelt down next to the one closest to the *Aurora*. "One of these mechs served as a command hub for the others," he explained after examining the mech. "I'm going to see……. No. Their batteries are all drained. They must have been out here digging in the dust for a long time. No matter. If we need a mech for something we can bring one in from the *Aurora*. My mechs are far superior to these anyway."

"What about the metal shards?" asked Eve. "The Alpha panel said it was consistent with a heavily damaged shuttle craft. Why would a shuttle craft have crashed here?"

"And the location of the shards doesn't make any sense," said Lin. "They look like they are all piled against the side of the Citadel. It wouldn't seem like that would be a natural distribution of debris from a crash, but I also can't imagine why anyone would have moved the debris there afterwards."

Jack was taking some readings from a scanning device on his wrist. "There is an abnormally large crater in that direction," he said pointing. "But it's not consistent with an impact crater. It was made by an explosion of some sort. And the shockwave of the explosion pushed the metal shards up against the Citadel." He was clearly very pleased with himself. "So the shuttle craft crashed first and the explosion happened after that. Maybe we will find out more once we are inside."

The trio began walking towards the rise containing the Citadel. As they moved along, they saw a variety of mechs in different positions. All of the mechs were inert and all were equipped with shovel-shaped implements. Jack's theory about how the face was created had to be correct.

The large door to the Citadel was closed. Jack opened it and stepped inside, still taking readings with his wrist scanner. "The power levels are very low," he said. "No maintenance likely for a few hundred ellipses at best. And if there haven't been any conscious Umae here during that time, the base would go to a reduced power setting. No need for environmental management other than that required to preserve the equipment."

"What about repose chambers?" asked Lin. Eve had the same question but hadn't been able to push it past her lips.

"The repose chambers get a priority energy feed," said Jack. "They would be one of the last functions to be shut down."

Inside the portal they saw a hallway that ended abruptly. Lin approached the wall and knocked on it. "Do we need to find another way in?"

"No," said Jack. He was examining the portal they had entered. "When this portal didn't close, the Citadel automatically sealed off this area to preserve the internal environment. With an exterior wall effectively breached all of the heat and oxygen inside would be lost too quickly. The environmental management systems would have overloaded eventually. Let me check something."

Jack moved over to a mini-panel along one wall and entered a series of directives. Once they were complete, the portal to the exterior slid shut. As soon as the portal closed, the wall next to Eve slid aside revealing an extended hallway that was dimly illuminated. On one side of the hallway was an auxiliary panel except this one also had a display. Jack moved to the panel and activated it.

"So this is the Citadel," mused Lin. "I hope it is more impressive when it is fully powered up."

"It will be," offered Jack as he manipulated the panel. "And I've never been here, I've only reviewed its schematics. Maybe after I do what I need to do on Uma, I can come up here and really implement some improvements."

"What are you finding?" asked Eve. There was a sharp edge of tension in her voice.

"I've had to activate the solar array," said Jack. "The fusion generators are offline. Once we store enough energy off of the grid, we can ramp them up. Then we will be able to see this place in its full glory. It shouldn't take too long. But good news. One live Umae is registering in the repose area."

Already the lights in the hallway were growing brighter. They could hear whirring sounds in the distance as dormant systems began to come back online.

"Is there a map?" asked Eve.

Jack nodded. "The Alpha panel is straight ahead. The repose chambers are off a side passage on the way. We are going to the Alpha panel first."

"No!" protested Eve. "I want to see Min! I need to see him!"

"I agree," said Lin. "I'd like to do a physical evaluation of him. Time could be critical."

"Time is critical," said Jack. "For the Umae. I need to know what happened down there before we do anything else. I need to get down there and deal with Dom. That takes priority over everything else. Once we've reviewed the Alpha panel, we can see about Min."

"Why can't we just go to the repose area by ourselves?" asked Lin. "It's not as if we will need you in there."

Jack scowled. "You need me everywhere," he snapped. "And I've already locked out the repose area's controls to everyone except me. Command Authority has its advantages. So you are going to obey me whether you want to or not!"

Eve stepped forward and attempted to pummel Jack's chest with her fists. "You can't....."

His reflexes were stunning. Eve's attack had met with partial success only because it never occurred to him that he would experience such disrespect from his underlings. His backhand caught Eve across the face of her helmet and sent her sprawling backwards to the floor.

"No!" Lin scrambled down next to Eve and quickly began assessing her. "You could have damaged her helmet! What were you thinking?"

"I'm thinking I'm the savior of our race!" screamed Jack. "And I know you agree! Now get her up and let's get moving. My people can't wait for me any longer!"

Jack walked briskly down the hallway towards the Citadel's Alpha panel. Lin and Eve hustled along behind him, fearing what he might do if they weren't watching. Just as they arrived at the panel, the area's illumination brightened considerably. The other displays flickered to life as their respective panels were activated by the increased energy provided by the generators. Jack turned and spread his arms wide. "No coincidence!" he proclaimed. "The Alpha panel has come back online just as I arrived. It is destiny revealed." He quickly referenced his wrist scanner. "The temperature and oxygen levels have been restored. We don't need these suits anymore."

He twisted off his helmet and set it on the floor before sliding out of his vacuum suit. Lin and Eve did the same. While the air smelled slightly stale, it was warm and comfortable. The Alpha panel finished with its initiation sequences and its display indicated its ready status. Jack wasted no time in entering inquiries about any significant events recorded by the Citadel via the beacon satellites since the **Aurora** had left for Haven.

"How did you determine your search parameters?" asked Eve. "Don't you think.....?"

"Quiet!" snapped Jack. "And I think much better than you do. Be still."

"Jack if you want information fast, you need my help. I don't care how smart you are, you aren't a Data Agent. You might miss something important."

"Nonsense," Jack muttered as he studied the information displayed in front of him. "But I do have a different problem."

"What's that?" asked Eve.

"There is quite a bit of information. I asked for a comprehensive update and the data is very...comprehensive. I could stand here reading through all of it, or......"

"Or what?" asked Eve. She had a pretty good idea where he was heading.

"You need to interface with the panel," said Jack flatly. "Then summarize for me the important parts of the story. If I think any important information is missing, I'll just ask you."

Lin took Eve by the elbow. "Eve, you don't have to do that. I don't have any reason to think that you are a high risk for additional trauma, but there's no good reason to take any risk at all. We've been gone for about a thousand ellipses. The rotation we would need to read over everything can't possibly make a difference now."

"Lin, shut up!" spat Jack. His voice oozed with hostility. For a moment she feared he might attack her. "That simply demonstrates that my people can't wait any longer than necessary. If you are interested in accessing Min while he's still alive, you'll do that interface. Now."

"You wouldn't dare!" began Lin.

"Try me," growled Jack. "Eve?"

Eve looked sheepishly at Lin before moving towards the panel. "What, specifically, do you want to know?"

"The beacon satellites transferred a great deal of data concerning some event that took place shortly after we left. There are also entries made by Min and by someone else who isn't identified. I need to know what happened and what Min and this other entity had to say about it."

Jack yielded his position in front of the panel so Eve could stand in his place. Lin stood back with her arms folded. She was seething.

Eve closed her eyes and lost herself in the process. If any of the prior conversations had affected her emotionally, it was no longer apparent. Lin marveled at Eve's ability to separate her feelings from the cold analysis she would be asking her mind to perform. After drawing one last deep breath, Eve placed her palms on the display and lowered her head.

Jack fidgeted. Any delay was almost unbearable for him. He stood by opening and closing his fists waiting for Eve to complete her interface. It didn't take long.

Eve stumbled back from the display, her eyes snapping wide open. Lin moved to her side. "Are you all right?" she asked quietly.

Eve nodded. "It's.....give me a moment." She closed her eyes and slowly shook her head. "All right. The beacon satellites executed a pulse wave. It spread over the entirety of Uma's surface. The energy from the wave converted much of the water in the atmosphere into rain. The entire planet experienced a constant rainstorm lasting nearly forty rotations."

"I already knew most of that," groused Jack. "That explains the planet's appearance. But why? What caused the pulse wave?"

Eve closed her eyes again, asking her mind to bring that information to her consciousness. "It was triggered by something Min did. He had information about the Hek's protein schematics. When he entered that data into the panel, the directive to execute the pulse wave was activated."

"A contingent directive?" mused Jack. "The aliens must have accessed the Citadel at some point and entered some of these directives. But why would they seek to destroy the planet simply because Min had information about the Hek's protein templates?"

"No." Eve's eyes were still closed and her voice was a mere whisper now. "The wave wasn't intended to destroy the planet. It was intended to effect Critical Closure. It would have rendered the Umae essentially sterile. Eventually, our numbers would have paled in comparison to the Hek's. The pulse wave was never intended to impact the atmosphere like it did. It was.....a punishment. Min had information he wasn't supposed to have."

"His possession of the Hek's protein schematics was going to be punished by exterminating the Umae?" asked Lin thinking aloud. "Why would the aliens care about that?"

"There must have been something about the Hek that was a threat to the aliens," said Jack. "Panel, Command Authority Command Agent Jack."

"Command Agent Jack recognized," responded the panel.

"Do you have the data I streamed to the Citadel through the tether beam from the *Aurora*?"

"Wait," said Lin. "You did what?"

"Affirmative," replied the panel.

"The Citadel collects all of the data sent back to Uma through the tether beam. I sent all the data you collected in the pyramid on Haven back here. To make sure I had it."

"You could have just asked....." began Lin.

"Panel, I want you to analyze the protein schematics of the Hek and compare its functions to the functions resulting from the aliens' protein schematics. Note any significant conflicts or augmentations resulting from their interactions."

"I'm not following you," said Lin. "What do the Hek and the aliens have to do with one another?"

The panel was rapidly processing Jack's request. The Journeyers could hear it humming at a slightly higher pitch.

"I looked at the protein schematics of the aliens back when we were still on Haven," said Jack. "They are essentially identical to the protein schematics Sol and I found on the **Aurora** when we first went aboard to try and determine what happened to her crew."

The panel was finished. "The Hek's neural structures would render them immune to the aliens' ability to influence other intelligent species with their audial signals."

"That's impossible," said Lin. "The two species originated from worlds hundreds of light ellipses apart. Such a characteristic could have never arisen through coincidence. It would have had to have been an evolutionary response. But that would mean......."

"That would mean that the aliens and the Hek were both on Uma a long, long time ago," said Jack, finishing her thought. "And let me lead you to the answer, since you won't find it yourself. Who built the Citadel and provided us with our technology?"

"The Directors did," said Eve. "But everything we know about them is clouded by time."

"Now we know one more thing about them," said Jack. "The aliens *are* the Directors."

The passageway was silent except for the vague sounds made by the Citadel's operating systems.

Eve finally shook her head. "No. It can't be."

"Yes," said Lin quietly. "It can. Jack's right. We've been their tools this entire time. The Umae, the Journeyers, all of us. Everyone except the Hek. The original brood pair came here at some distant point in the past and

217

found the Hek. The Hek were supposed to be their proxy species. As the Hek developed technologically, they would eventually create the means to radically alter Uma's atmosphere so it could become a new brood world. But something happened along the way. The Hek evolved and developed a resistance to their control. That explains the geographical barriers between the Hek and the Umae."

"Yes," agreed Jack. "An excellent connection. Actually one I hadn't considered. The aliens must have separated the Hek who had not evolved to resist them from those who had. This first group was isolated from the second. Islands, mountain ranges, lakes, deserts. I guarantee that if we mapped out all of the former locations where the Umae had settled, we would find natural barriers of all sorts keeping them away from the Hek. No doubt they hoped that the Hek who could resist would simply die off. But they didn't."

Eve seemed to be doing calculations in her head. "We descended from the Hek," she concluded. "And we were to become the aliens' proxy species. Imagine the planning. Once our language developed sufficiently, they offered us the story of the Directors. And through the Directors they controlled everything. They vilified the Hek so we would fear them. They gave us a baseline of high technology just advanced enough to enable us to eventually redesign the atmosphere. For them. And they used the technology gap between the Journeyers and the rest of the Umae to set the Journeyers up as figures of worship." She shuddered slightly. "We were the foundation for everything. And it was all for them."

"Panel what is the current population of Uma?" asked Jack. "Umae and Hek?"

The panel hummed again as it drew information from the beacon satellites. "There are approximately 100,000 Umae. The Hek have been extinct for nine hundred and six ellipses."

"Extinct?" repeated Lin. "How?"

Eve closed her eyes, her face again growing calm. "The pulse wave. It rendered them infertile. They could not reproduce."

"But I thought the pulse wave was designed to sterilize the Umae?" asked Lin. "Was it flawed?"

Eve shook her head. "No, it was changed before it could impact Uma."

"By Min," said Jack. "Those energy bubbles you mentioned before, Eve. That's how he did it. He changed the wave so it would sterilize the Hek instead of the Umae."

"But there were two bubbles," said Lin. "At least that is what Eve said earlier. Eve?"

"Yes, two. The second was created by the other entity."

"Panel, did Min calculate the effect of the wave with only his energy bubble?"

"Yes," said the panel. "Effective sterilization of the Hek with moderate impact on the planet's environment."

"And the impact of the second bubble?"

"A greater probability of sterilization and significant damage consistent to its condition now."

Jack nodded. "As I thought. The other entity was the missing brood queen. She came here on the *Aurora*'s shuttle craft. She must have controlled one of her crew and was already here when Min arrived."

"Ask the panel if there are signs of the aliens' protein schematics within the Umae genotype," asked Lin.

"Panel?" asked Jack.

"Approximately half of the Umae bear alien protein schematics," said the panel. "It is impossible to tell based on current available data how much of an Umae genotype has to be composed of alien schematics before that individual would act in the aliens' interests. The highest density of alien proteins is roughly centered over one geographic area on the planet's surface."

Jack smiled. "That's where we will find Dom."

"What about the brood queen?" asked Lin. "Aren't you worried about her?"

Jack shook his head. "She's dead. Min killed her. I'm willing to bet the future of the Umae species that the blast crater we saw outside was where she died. We will have to ask Min when we wake him up."

"Otherwise there wouldn't be any Umae left on Uma," noted Lin.

"Exactly."

"There's something I don't understand," said Eve. "Was there some other way Min could have stopped the pulse wave from sterilizing the Hek? Anything."

"Panel? Anything?" asked Jack.

"Affirmative. A relatively simple adjustment to the first energy bubble would have deflected the wave out into space leaving Uma and its inhabitants entirely unaffected."

"Why then?" said Eve somberly. "Why did Min sterilize them when he didn't have to? Why didn't he just deflect the wave into space?"

"Another good question," said Jack. "You can ask him that too."

The three stood quietly next to the panel, each searching for alternative explanations or new questions they hadn't thought of. None of them spoke until Jack's anxious energy finally returned.

"Let's go wake up Min," he said finally. "He will have some answers we need. And then? Then I'm going to go save our planet."

The Citadel's repose area had been a goal for all three of the remaining Journeyers. Jack was determined to show Min the extent to which Jack's grasp of technology now surpassed his teacher's. Lin was eager to implement the medication she had developed for RTS. Although Jack's plans would possibly render future space travel unlikely, it was still a breakthrough she believed would greatly benefit the Umae. At a minimum, she could likely save Min.

Eve's entry into the repose area was far more emotional than she could have imagined. Min hung limp within the anti-grav field generated by the chamber. Although Lin had assured her that he was still alive, his gaunt, lifeless form sent tremors through her chest. She ran to the chamber and pressed her hands against its transparent walls. Next to Min's repose chamber, a skeletal form was sprawled on the floor. It had a noticeable hole in the center of its forehead.

"Lin?"

"One moment," said the Life Agent quietly. She activated the mini panel on the side of the chamber and reviewed the data it provided. "He's alive. He's been in repose for a long, long time though. His systems are extremely compromised." She drew a deep breath and released it slowly before turning towards Eve. "I don't think we should bring him out."

Eve's face was overrun with panic. "What? But.....no! That can't be right. Jack?"

Jack stepped up next to Lin and studied the data himself. "If we try to bring him out, he will likely die," he said finally. "But so what? If we just leave him in there, he may as well be dead anyway. Let's do it."

Lin stepped in front of him. "No! I won't let you!"

The gigantic Command Agent peered down at her and started laughing. "And what are you going to do to stop me?"

Lin glanced back and forth between Eve and Jack. Eve was beyond distraught and her eyes were focused solely on Min. "You want those answers, right?" she asked finally. "About the brood queen? About the deflection of the energy pulse?"

"He's not going to give us those answers any time soon, is he?" sneered Jack.

"He could," said Lin looking back at Eve. Eve finally tore her eyes away and returned Lin's glance.

"Yes. He could," said Eve brightening. "He can give them to me."

"What are you talking about?" demanded Jack. "This is nonsense."

"No, it isn't," said Eve. "If I understand Lin correctly, bringing Min out of repose, even with her medication, will likely result in his death. But I can interface with him. I can communicate with him without waking him up."

Jack paused. "It's possible, although I'm not aware of any Umae interfacing directly with one another before. Umae aren't machines, at least not in this sense."

"I wouldn't be interfacing directly with Min," explained Eve. "His nervous system is being monitored by the repose tube. I could interface with him using that connection as a conduit. It would be like communicating with him through a dream."

"You've done this before?" shot Jack. "You've ever *heard* of this before?"

Eve shook her head. "Only in theory," she admitted.

"Eve's already done it, at least on the receiving end. Lise interfaced with Eve on the **Starshine**. Eve just can't remember it. And," added Lin, "the risk is minimal. Even if it doesn't work, neither of them would be endangered."

Jack sighed with resignation. "How do you actually effect the interface? The tube isn't designed for that."

"No, but the panel that monitors Min's nervous system would be the same one that monitors the nervous systems of everyone in repose on the Citadel," said Lin. "The tubes are already connected in a sense. As I understand the process, Eve would have to somehow 'find' Min."

"Yes, your understanding is correct," said Eve. "Through the conduit of the panel."

The repose area had over twenty repose tubes. They were all lined up against the same wall.

"I'll make it easier for you," said Jack. "Since we only need two tubes for this to work, no sense in confusing Eve with extras." Jack began moving from tube to tube, separating each from their connection to the repose chamber's primary panel. Soon the only active tubes were the one Min was in and the one adjacent to it.

"I'll monitor you from the tube's mini panel," said Lin to Eve. "But you are going to be fine."

221

"And remember," said Jack, "I need to know whether or not the brood queen is still alive. Because if she is, I'm going to have to change my plans for Dom. If you want to ask him anything else, feel free. Just don't take too long."

Eve nodded and stepped into the empty chamber. Lin made sure all of the connections were proper before gently touching Eve on the cheek. "You'll be fine," she said quietly. She stepped back and let the door to the tube swing shut and then watched as Eve slid off into repose.

"How long will this take?" demanded Jack.

Lin shrugged as she eyed the mini panel. "Impossible to say. To my knowledge no one has actually attempted a tube-to-tube interface. But Eve has a couple of things working in her favor."

"Oh? Like what?"

"One, she is an unbelievably talented Data Agent. And two," she added in a hushed voice, "she is very familiar with her interface target."

Jack poked at the skeleton with his foot. "This poor fool must have been on the *Aurora*. He got on the wrong side of the brood queen, I'd guess."

Lin did not look away from the mini panel. "Everything appears nor-...... no, wait!"

Jack nearly bowled her over as he stepped up to look at the display. "What is it?"

Lin forced her way back in front of the display. "Her heart rate is escalating, the neurotransmitter levels in her brain are spiking......"

"What's wrong with her?" asked Jack. "We should get her out of there." Eve grabbed his arm with both hands as he reached for the tube's controls.

"No! Don't do that! If we force her out there will be a modest feedback loop through the entire system. It won't hurt her because she is healthy but Min might not be able to handle it."

"But she's in trouble!"

"No, she isn't," said Lin with a smile. She beamed as the levels of certain neurotransmitters in Eve's brain soared. "She's in love." She released Jack's arm.

"So, she's found him."

"Yes, as I knew she would. After all of this time she has found him. Again. It shouldn't be long now."

Eve's brainwaves shifted slightly prompting Lin to deactivate the tube. Within moments, Eve had regained consciousness. She stepped from the tube into Lin's awaiting arms, her face radiant.

"Oh my stars!" she gushed. "He's alive! He's alive! I never thought......"
She buried her face in Lin's shoulder as her body shook with sobs.

Jack fidgeted impatiently. "Let's move on. What did you find out?"

Eve stepped away from Lin and struggled to compose herself. "You were right. The brood queen is dead. Min tricked it into blowing itself up in the shuttle. It was the last thing he managed to do before he went into repose."

Jack nodded. "Excellent. That will make things easier for me. All right, let's get out of here."

"Wait!" said Eve. She turned towards Lin. "He didn't know about the Hek's immunity to the aliens' manipulation ability when he decided to sterilize them. He didn't know. And the second bubble was created by the brood queen."

Lin's forehead creased with confusion. "But why then? Why didn't he just deflect the wave into space? Why did he decide to sterilize the Hek?"

Eve's body shook with a renewed wave of sobs. She put her hands gently on Lin's shoulders. "He was trying to prevent the possibility of Critical Closure from ever taking place. He did it because........" Eve tried to draw a deep breath as her shaking intensified. Finally she managed to whisper into Lin's ear. "He wanted to save the small chance that he could still be with me."

Lin closed her eyes as a lone tear rolled down her cheek.

"I have one more thing I need to ask him," said Eve. The women's eyes locked as they joined hands. "Please." Lin gave Eve's hands a firm squeeze.

"No, there's no time!" complained Jack.

"Go," said Lin as she began preparing the tube again. She glared at Jack. "If you need me for any part of your plans concerning Dom, you are just going to have to be patient."

Jack returned her stare before turning away.

"Thank you," said Eve with a sniffle. She gave Lin a long, warm embrace.

"Don't tell him about what happened to the Hek," said Lin. "There is no reason for him to bear that now. Talk about...happy things."

"Yes. Happy things." Eve smiled at Lin, a fresh reserve of calm flowing over her. "You happened with me," she said quietly.

Eve stepped back into the chamber. Lin reactivated it and Eve began a new repose cycle.

"How long?" groused Jack. "I need to go!"

"We can go now," said Lin as she began walking towards the exit.

"But what about them?" said Jack pointing at the chamber. "Didn't she say she had something else to ask him?"

"She did," said Lin. "But that's between the two of them. She's not coming back out. You don't need her anymore. Let her go."

Jack frowned before turning and stomping past Lin. Lin walked slowly to the exit before stopping and turning around. Min and Eve were floating next to each other in adjacent repose tubes, connected by an electronic interface. If nothing else turned out right for Lin, at least she would always have this.

"And you, my dear friend, happened with me."

CHAPTER 22

Lin wasn't sure if Jack would wait for her to get to the *Aurora* or not. The fact that he did confirmed her belief that he expected her to play some role in his confrontation with Dom. By the time she boarded the ship, he had already shed his vacuum suit and was standing next to the door to his personal area.

"Get to the Life Lab and run the bio-protocols for re-entry," he said. "It would be tragic if the Umae all died from some extra-planetary microbe after I managed to save them from extinction."

"What are you going to be doing?"

"Getting ready. After we clear what's left of the atmosphere, I'll tell you exactly what you are to do. And you will do it exactly."

"Do I get any hints?"

Her question was wasted as Jack had already disappeared into his room. She tossed her vacuum suit to the side, not bothering to properly stow it away. She made the long walk back to the Life Lab to begin her re-entry protocols. There was the gap where Sol's repose tube had been. She paused to reminisce on all the lessons the old Life Agent had shared with her. His care and concern for each of his patients was surpassed only by his care and concern for the Umae in general. Where she expected grief, she found only the same comforting warmth she had always experienced when she was with him. He had happened with her and would continue to do so for as long as she lived.

Frell was as timeless as ever. She looked upon him for longer than she had intended, enthralled by the depth of his sleeping features. Although she realized he would change very little while he was in repose, it occurred to her that he had retained a certain consistency to his nature from the first time she met him as a child. Wide-eyed, smiling, quirky, full of surprises. He defied expectations. His unpredictability was a constant. When Sol had expressed his opinion about Frell's potential, Lin was shaken. But as she had had the chance to think about it since then, it felt right. He seemed quirky and unpredictable because he was different. His difference was in how he perceived his world and processed its possibilities. He was better. Sol had seen what others had not.

As she sat down at the Life Lab's Alpha panel, Jack's voice came over the com-link.

"We are ready to enter the atmosphere," he said. "Where are you on your bio-protocols?"

Lin grimaced. "I just got here. Can you slow down?"

"No!" he cried. "I can't! You don't seem to understand how badly my people need me. Just run a quick biomass analysis instead and get back up here. I need you."

Lin reached out and turned off the com-link. Jack could turn it back on with his Command Authority but she was guessing he wouldn't bother.

"Panel, I need a routine biomass analysis of the entire ship," she said glumly. She knew what the result would be. The odds of a Journeycraft transporting some deadly microbe from outer space were astronomical. But then she recalled how the brood queen had managed a ride on the *Aurora*. It must have evaded the biomass testing by dropping off before the crew prepared to land. The biomass analysis would initially analyze the living matter on the ship to see if any of several characteristics exceeded a specific standard range based on known life forms and those expected to inhabit the ship. The assumption was that any measurement beyond that standard would be the result of something foreign.

"Biomass exceeds standard range," replied the panel.

"What? Explain."

"The overall biomass on board the *Aurora* is approximately twenty-six percent beyond the expected range."

Lin groaned. "Of course. Recalibrate to account for Command Agent Jack's increased mass."

"That was already factored into the analysis."

Lin frowned. Unless the panel was malfunctioning, something with a substantial mass was alive and on board the ship, or attached to the ship, that she didn't know about. She reactivated the com-link.

"Jack? The panel is reading an excessive biomass. Care to tell me what you're growing in your secret room? Jack?"

The com-link remained silent. Did the panel have the ability to scan the inside of Jack's chamber?

"Describe the organisms responsible for the additional mass other than Umae passengers."

"Umae passengers comprise over ninety-nine percent of the biomass."

She grimaced. Jack's fabulous ship was choosing a fantastic time to malfunction.

"Review baseline assumptions." Lin had reduced the crew complement by one to account for Eve's absence for the determination of the biomass standard. There was a small possibility the panel hadn't accounted for the change.

"Baseline assumptions include the fact that there are three Umae on board, one being a child."

Lin found an odd pleasure in the fact that Jack's panel was having such a difficult time performing a routine function. "Well it isn't like we picked up an extra crew......"

She froze. Jack had insisted on taking the time for a full bio-protocol analysis but quickly changed his mind to the much simpler biomass measurement. He could have had the ship's Alpha panel do that without her. In her frustration and eagerness to be away from him that hadn't occurred to her. Her chest was encased with dread.

"How many distinct Umae are included in the biomass?" she screamed.

"Four."

Lin exploded out of her chair and sprinted out of the Life Lab back in the direction of the bridge. "Jack! Jack!" She was screaming as loudly as her lungs allowed. As she arrived on the bridge, she saw Jack standing next to a portal staring out at the planet below. She heard a muffled whoosh from inside the walls of the ship.

"You are just in time," he said without turning around.

Lin dashed to an adjacent portal and looked out. Below she saw a dim streak of light as an object moved quickly away from the *Aurora*. A glow started at its nose and spread to the body of the object. Then it burst into a spectacular display of bright green.

The colors formed spots in Lin's vision. "What? What did you do?" she screamed as she ran at Jack.

He effortlessly seized her by the wrists as she approached. "We all had a role to play," he said matter-of-factly.

The com-link on the bridge activated. "Biomass excess corrected. Biomass within safety parameters."

Lin pulled as hard as she could in an effort to free herself from Jack's grip. "You killed him! You killed Sol!" she screamed.

"He was as good as dead," explained Jack without letting go. "I needed a way to get Dom's attention. And, after all, he was due a hero's farewell. His was much better than the one Min gave Hab."

Lin's strength failed as her knees buckled and she slid to the floor. She rested her head on the cold metal deck. She could feel the vibration of the landing boosters initiating. She forced herself back to her feet and sat in the nearest chair. Jack appeared to be moving in slow motion as he manipulated the Alpha panel. "What now?" she said in a flat voice.

"Stay here," commanded Jack. He re-entered his chamber and closed the door.

Lin felt the ship touch down. No doubt Jack had used the *Aurora*'s scanners to determine the density spread of alien protein structures. Dom would be near its center. Sol was nothing more than bait for whatever Jack had planned.

She heard his chamber open and Jack clambered back out. He had donned a metal suit of some sort. She could see energy sparking through what appeared to be veins in its structure. Although she couldn't see any joints in the fabrication, Jack moved fluidly. This was his secret project. He strode easily towards the egress ramp and beckoned her to follow.

Once they reached the bottom of the ramp, he turned towards her. In his hand was a metallic wand about as long as her arm. On one end was a shiny ball and near the handle was a simple red button. After he handed it to her she had to grip it with both hands to avoid dropping it.

"What is this?" she asked listlessly.

"I'm going to kill Dom after he finds me," said Jack. "His arrogance will be his undoing. It won't occur to him that I bring his doom. But I need to be sure that he finds me before he finds you."

"What do you mean?"

"Above all, Dom must not access the *Aurora*. He wouldn't be able to access any of the systems because I have secured them, but he might be able to damage the ship. It is the key to the post-Dom world I plan on building for the Umae under my guidance."

His voice, although cool and metallic, dripped with mania.

"Tell me what to do."

"You must maintain the security of the *Aurora*. If Dom finds you here before he finds me, use this. It is a weaponized focus reactor. It will emit a massive energy burst that will certainly kill him outright. So regardless of who he finds first, his death will be at my hands."

"Then what?" she hefted the wand, its weight taking a toll on her arms.

"Then the Umae are restructured by my hand!" shrieked Jack. "Greater than ever before, as have been all of my modifications. The *Aurora* will enable me to subject the Umae to the grandest improvements. As a species, they will reign supreme. And I will guide them."

Lin knelt down, setting the butt of the weapon on the ground so she could rest her arms. "I'm eager to serve," she said passively. Jack turned on his heel and moved off towards a nearby forest, his suit enabling him to make enormous strides.

Lin looked down and saw blood dripping from her hands. She had been digging her fingernails into her palms. She set the wand against the side of the egress ramp and re-entered the ship. She saw her equipment bag on a nearby table. Her Tissue Stimulator would quickly stop her bleeding. Digging through her bag, she cursed herself for not being more organized. Her fingers ran across a hard, smooth surface. She withdrew the Chromatic Striation Chamber Sol had given Frell. Where it had formerly been filled by a liquid with rainbow-colored striations, it was now filled with........goo. Frell had solved the puzzle of the chamber. The contents of the tube were his reward.

Lin set the chamber aside briefly as she located her Tissue Stimulator. A brief application stopped the bleeding from both of her hands. She then walked over to Frell's repose tube. One remarkable individual had the ability to shape the direction of his people for generations to come. Quirky, unpredictable, silent. Brilliant.

Jack's armor included a proximity scanner among its broad variety of devices. It enabled him to detect movement of certain types in a one-hundred-and-eighty-degree arc directly in front of him. He had designed this one to specifically detect the aliens' protein templates. There was only one target he was interested in encountering. Dom. And it didn't take too long before Jack's scanner found him.

Jack laughed to himself inside his armor. Of all his technology, his battle suit was his crowning achievement. The only possible device that matched its ingenuity was the reactor wand he had given Lin. Lin was a dullard. It was imperative that she be empowered to deal with Dom if necessary. Jack's plan would be much more difficult to accomplish without the *Aurora*.

He could have simply killed Dom from space, but that wouldn't have been nearly as satisfying as killing him with a technological creation entirely his own. While Jack had improved the *Aurora* far beyond the limits of any Journeycraft ever constructed, it was still, at least in part, a creation of others. Dom's death would be Jack's glory to claim and Jack's alone. And now that Jack had Dom on his scanner, he could keep himself between Dom and the *Aurora*. Nothing could save Dom now.

As Dom entered the clearing he smiled broadly at the armored figure standing in his path. He couldn't see the face of the gigantic creature within the suit, but it didn't matter. There was a Journeycraft nearby and it was the tool through which Dom would find his personal salvation. He continued to advance. Don raised his hand bidding his warriors to halt.

The armored figure moved forward as well. "I was right, as usual," said Jack. "I have completely figured you out, Dom. And your defeat is imminent."

Dom shrugged as he looked up at the giant's head. "You have me at a disadvantage," he said. "You know my name, but I don't know yours."

Jack stopped. "I'm sure you will remember me. I am Command Agent Jack of the Journeycraft *Aurora*. It is only fair that you know the name of the one who ended your reign."

"Jack? You have grown quite a bit since the last time I saw you. You were Min's snot-nosed errand boy what, several hundred ellipses ago? I'd hoped you might return some rotation."

"You have been wishing for your doom then," said Jack confidently. "Min killed your queen, and I'm much more powerful than he ever dreamt he could be. Your end is at hand. I'm thrilled that you brought your genetically amplified lackeys with you. Now I won't have to hunt them down."

"It sounds like you have big plans for the future," noted Dom.

"Indeed," said Jack as he began adjusting the panel on his chest. "I'm going to elevate the Umae to a status of excellence far beyond anything your pathetic species has seen. And I will be at the apex." He completed his adjustments. "There. I don't intend to be cruel. I assure you that your death will be swift and painless. Mostly."

Jack was taken by the return of Dom's confidence. A moment prior Jack was sure he had noted fear in the voice of the alien-child. That fear vanished as a broad smile swept across Dom's face.

"Perhaps one day," he said cheerily. "But not today." His voice dropped several octaves and became a slow flow of cold speech. "I know you want to join me, don't you?" His voice dripped with shadow. "Serve at my side."

Jack burst into laughter. "Fool! I know more about you than you know about yourself! Your manipulations can't work on me!"

"I wasn't talking to you," said Dom calmly. Again, he utilized his prior voice. "Kill him."

Jack stopped laughing for a split second before his entire form was engulfed in a flood of crackling light. Lin stood behind him gripping the wand as an energy beam poured from its end. Jack's body shook as he slid to the ground, his form writhing under the assault. Lin continued focusing the beam on him, her arms shaking from the effort. Finally, she stopped and dropped to one knee, panting heavily. Jack's form was prone on the ground emitting a foul-smelling smoke. He was struggling aimlessly. Initially, he appeared to be trying to rise but then listed back to one side.

Dom was beside himself. He doubled over from uncontrollable laughter. These Umae, even the Journeyers, were so stupid. His soldiers began moving up behind him. Dom straightened himself and waved Lin forward. "Come cow! We will all use your pathetic form as our tool of celebration!"

His body snapped upright as his body shook under a new assault from Lin's wand. His muscles were locked in place by the energy coursing through his limbs. The smell of his own burning flesh filled his nostrils as his vision grew gray. With a final primal cry, he collapsed next to Jack. Lin continued to focus the energy beam until her shaking arms failed and she dropped the wand. She dropped to her knees and gasped for air. Dom's minions glanced back and forth at one another, unsure as to what to do.

Lin reached into each ear and dug out a glob of multi-colored goo.

"Woman!" cried out one finally. "You shall taste the full force of our wrath!"

The wand was far too hot for Lin to grasp. She crawled forward until she reached Jack. The face plate of his armor was burned away and she could make out one functioning eye beneath the melted metal and flesh.

"*Aurora* ....is......mine," he said weakly. "Bluff."

She felt certain it was his last breath, so there was no time for questions. She stood up and stared defiantly at the approaching Umae, raising her hands high above her head.

"Woman!" the new leader hooted again.

A fan of bright orange spread into the sky over the forest. This light was chased by a deep rumble that rolled towards the field from behind her. Lin shielded her eyes with her hands and then looked away entirely once she realized it might endanger her vision. The Umae froze in place as they beheld this awesome power. Broken, they turned and fled in panic as the ground began to shake. Exhaustion finally claimed Lin. She stretched out on the ground, no longer caring what might happen to her.

When she awakened, it was dark. The light was gone and both Dom and Jack were motionless smoking heaps. The area was heavy with the horrid smell of burning flesh. Dom's Umae were gone. The only sound was the wind as it moved gently through the nearby trees.

Lin stumbled through the darkness, a tiny light from her equipment bag her only means of combating the night. She eventually saw trees felled as if by a sudden burst of wind. Farther on trees and bushes smoldered creating a hazy half-lit air. Then she reached her destination.

A broad, scorched crater stood where the *Aurora* had been. Jack had claimed it for himself. She dropped to all fours and looked down into the darkness. A raw heat was all that greeted her from the bottomless blackness before her. She rose and moved to the very edge of the pit, closing her eyes, and remembering the little boy who saw a world no one else could see.

EPILOGUE

The fire was dying again. The woman took the last of her wood and added it to the waning flame. Strands of gray-white hair escaped the thong she had tied to contain it. Her knees and back ached as she leaned in and fed the fire.

"Mother?!?" Someone was calling to her.

She could just barely see the mouth of her cave. She squinted to see if she could discern the source of the cry.

"Mother! Come!"

With a weary sigh, she stiffly reached for a burning branch to guide her way in the darkness. As she shuffled forward, she heard a scrambling noise in the nearby brush. Then she heard a loud bawling right in front of her.

Taking a few more steps forward, she peered into the shadows. She spotted a small wrapping wiggling on the ground.

"Moons," she muttered under her breath. Footfalls moved away from her into the deep wood. She couldn't tell who had made them. She never could. She walked to the bundle on the ground and drew back the top layer.

A baby looked up at her and let loose with a hungry howl. He was sturdily built and had light brown hair. His eyes were the same shade of blue as the last one's had been. He grabbed the strand of hair that had escaped her thong and pulled. She pried the baby's fingers away, causing it to cry out again.

The wood contained nothing but the wind now. She and the baby were alone. She lifted him into her arms and began walking back towards her cave.

The baby, like the others before him, was past weaning. Perhaps the simple concoction she had cooking on her fire would suffice to quell his hunger for now.

Lin shifted the infant onto her hip. In the morning, she would check her traps for food.